Genomics of GC-Rich Gram-Positive Bacteria

Functional Genomics Series Volume 2

www.functional-genomics.org

Edited by

Antoine Danchin
HKU-Pasteur Research Centre, Hong Kong

Series Editor: **Milton H. Saier, Jr.**
University of California at San Diego
La Jolla, USA

CAISTER
ACADEMIC PRESS

Caister Academic Press
32 Hewitts Lane
Wymondham
Norfolk NR18 0JA
U.K.

www.caister.com

British Library Cataloguing-in-Publication Data

A catalogue record for this book is available from the British Library

ISBN: 0-9542464-3-8

Printed and bound in Great Britain

Contents

Contributors

Yanis Ben Amor
Public Health Research Institute
225 Warren Street
Newark, NJ 07103, USA

Patrick J. Brennan
Department of Microbiology
Colorado State University
Fort Collins
Colorado 80523, USA

Jean-Christophe Camus
Institut Pasteur,
28 rue du Docteur Roux
75724 Paris Cedex 15, France

Stewart T. Cole
Institut Pasteur,
28 rue du Docteur Roux
75724 Paris Cedex 15, France

Antoine Danchin
HKU-Pasteur Research Centre
Dexter HC Man Building 8
Sassoon Road
Pokfulam, Hong Kong

Maria Laura Gennaro
Public Health Research Institute
225 Warren Street
Newark, NJ 07103, USA

Brigitte Gicquel
Institut Pasteur
Unité de Génétique Mycobactérienne
28, rue du Dr Roux
75015 Paris, France

David Hopwood
John Innes Centre
Colney
Norwich, NR4 7UH, UK

Louis M. Jones
Institut Pasteur,
28 rue du Docteur Roux
75724 Paris Cedex 15, France

Leonid A. Kulakov
The Questor Centre
David Keir Building
The Queen's University of Belfast
Belfast BT9 5AG,
Northern Ireland

Michael J. Larkin
School of Biology and Biochemistry
Medical Biology Centre
The Queen's University of Belfast
Belfast BT9 7BL, Northern Ireland

Philippe Mazodier
Unité de Biochimie Microbienne
Institut Pasteur
25 rue du Dr Roux
75724 Paris Cedex 15, France

Ivan Moszer
Institut Pasteur,
28 rue du Docteur Roux
75724 Paris Cedex 15, France

Olivier Neyrolles
Institut Pasteur
Unité de Génétique Mycobactérienne
28, rue du Dr Roux
75015 Paris, France

Jean-Luc Pernodet
Institut de Génétique et Microbiologie
UMR. CNRS 8621
Bât. 400 Université Paris-Sud
F91405 Orsay Cedex, France

Melinda J. Pryor
Institut Pasteur,
28 rue du Docteur Roux
75724 Paris Cedex 15, France

Varalakshmi D. Vissa
Department of Microbiology
Colorado State University
Fort Collins
Colorado 80523, USA

Foreword

David Hopwood

When I graduated in botany from the University of Cambridge in 1954 I was looking for a challenging topic to study for my Ph. D., preferably in microbial genetics, which seemed to be on the verge of an explosion. Bacterial genetics was in its infancy, but had already thrown up some bizarre discoveries. Genetic recombination resulted from the transfer of packages of genes as naked DNA, or via the agency of viruses, or by a strange form of incomplete mating. In contrast, fungi and other microbial eukaryotes (as we later came to know them!) seemed to generate recombinant genotypes essentially in the same way as higher plants and animals, with an ordered alternation of haploid and diploid generations separated by complete nuclear fusion and meiosis. So, when Lewis Frost and Harold Whitehouse pointed out that the actinomycetes were regarded as intermediate between bacteria and fungi, and should therefore be a happy hunting ground for novel genetic phenomena, I took on the challenge. After a false start, I adopted a strain of *Streptomyces coelicolor* acquired serendipitously (Hopwood, 1999). My first rudimentary genetic map of the A3(2) strain, in 1958, had just six genes, arranged in two linkage groups. By 1965 the map (though not necessarily the chromosome) had become a single circle, the second example in a bacterium, with 40 genes. The map assumed physical form in 1992, largely through the efforts of Helen Kieser, and in 1993 broke established precedent by revealing itself to be linear, via the seminal experiments of Carton Chen and his colleagues. By 1996, the heroic efforts of a group spearheaded by Matthias Redenbach culminated in an almost complete set of cosmids covering the chromosome, and carrying some 170 genes mapped by Helen Kieser. The resulting high-resolution genetic/physical map, along with the cosmids, turned out to be a marvellous resource for the worldwide *Streptomyces* community, and also provided the jumping-off point for the Sanger Institute project to sequence the genome. This effort culminated in July 2001 with the identification and preliminary annotation of 7825 protein-coding genes on a chromosome of 8,667,507 base pairs of DNA. This is the largest number of genes in a microbial genome completed to date. An account of the main features of the genome appeared in May 2002 (Bentley *et al.*, 2002).

The actinomycetes turned out not to be intermediate between bacteria and fungi: their mycelial form, which confused many early microbiologists, evolved independently from that of moulds. They are eubacteria, but they are separated by enormous periods of evolutionary time from the simple rod-shaped bacteria where genetic observations began. The *Streptomyces* genome has indeed turned up a fair share of novelties, with its linearity, its unusual system of end-patching, and its arm regions devoid of essential genes and subject to extensive structural changes, including deletions, amplifications and horizontal transfer. Its information content is huge for a microorganism, probably reflecting an extreme flexibility in its responses to physical, chemical and biotic stresses in its chosen soil habitat. With thousands of 'orphan' genes, the challenge for functional genomics is huge, but the tools are being sharpened and the pay-off will be that much larger than for a more specialised organism with a smaller genome. As the analysis continues, we shall learn a lot about the genetic basis for a successful life in the soil. Humankind will also benefit from the enhanced possibilities that emerge for the generation of novel therapeutic agents by using the secondary metabolic gene clusters of the streptomycetes to do novel chemistry through genetics, and optimising *Streptomyces* expression hosts to make the resulting 'unnatural natural products'.

Jean-Luc Pernodet and Philippe Mazodier have written a highly readable account of all these aspects of the *Streptomyces* genome and genetics. It is particularly useful that their chapter is included in a book that also deals with the genetics of some of the actinomycete cousins of *Streptomyces*. Of these, most people would regard the Mycobacteria as the most important because members of this genus, *Mycobacterium tuberculosis* and *M. leprae,* cause two of the most dreaded scourges of mankind, tuberculosis and leprosy. The complete genome sequences of these two pathogens, completed in 1998 and 2001 respectively, have already proved to be a mine of information. The genome of *M. tuberculosis* contains fewer genes than *S. coelicolor* – almost exactly half as many – not surprisingly because it is clearly specialised for a much narrower ecological niche in human macrophages, while still being capable of (slow) growth on artificial media. Again, the finding of an even smaller genome in *M. leprae* came as no surprise, since it would seem to be even more specialised, as well as having apparently lost the ability to grow in culture. Few would have guessed, though, that this obligate intracellular pathogen would have seen fit to downsize its coding capacity to a mere 1600 genes! There is a wealth of information about these two pathogens in this book, including discussion of early conclusions about pathogenicity stemming from the genome sequences and a generically valuable account of the diverse protein secretion systems in *M. tuberculosis.*

There is also a detailed account of TubercuList, the database modelled by the group at the Pasteur Institute on their pioneering and public-spirited development of Colibri and SubtiList for *Escherichia coli* and *Bacillus subtilis* respectively. This user-friendly format is increasingly being adopted to present bacterial genomic data. As well as Leproma for *M. leprae*, Sco-DB II for the *S. coelicolor* genome has recently been launched (the latter supersedes Sco-DB, an earlier database in AceDB format).

The genetics of the actinomycetes has come a very long way since the days when they were thought to be intermediate between bacteria and fungi. At least in more recent years developments have kept up with the frantic pace of data acquisition on the genomics of bacteria as a whole. As well as the species dealt with in detail in this book, the genome of *Mycobacterium bovis* has been completed at the Sanger Institute, allowing highly informative comparisons to be made between the genomes of members of the *M. tuberculosis* complex, including two very different isolates of *M. tuberculosis*. The chromosomes of *M. avium, M. smegmatis, M. paratuberculosis* and *M. marinum* are also on the way, so pretty soon the comparative genomics of *Mycobacterium* will be a paradigm for bacteria. The much smaller genome of another actinomycete pathogen, *Corynebacterium diphtheriae,* has also been sequenced. An informative comparison involving this genome has already revealed fascinating glimpses into actinomycete evolution. The Sanger Institute's ACT (Artemis Comparison Tool) software revealed a significant synteny (conserved gene arrangement) between the *M. tuberculosis* and *C. diphtheriae* genomes, reflecting descent from a common ancestor. Each of these genomes showed a perceptible synteny with the *S. coelicolor* chromosome, but only with its core; the *Streptomyces* arm regions were not involved (Bentley *et al*., 2002). This could imply that the arms have been acquired by horizontal gene transfer since divergence from the last common ancestor of the pathogens and the saprophytic *Streptomyces.*

For the non-pathogens, once the genome sequences of other *Streptomyces* species are completed and, hopefully, released, most informative comparisons will become possible. For example, will the arm regions of different species be found to differ more extensively in gene content and arrangement than the core regions, as the hypothesis of horizontal gene transfer would predict? Meanwhile, too, another group of beneficial actinomycetes, the Rhodococci, will soon have a sequenced representative. The Genome Sciences Centre in Vancouver is sequencing the genome of a species of *Rhodococcus*, a potent agent of bioremediation and biotransformation. A chapter in the book tells us about the background that will be illuminated once this genome is completed.

So, for all of us fascinated by the actinomycetes, like anyone involved in prokaryotic research, we are entering a new Golden Age in which fresh knowledge about bacterial genomes is arriving with breathtaking speed. This would be a good time to re-visit some of Waksman's classical writings, as I have recently done, to realise on the one hand how far our understanding has come since his classic monograph (Waksman, 1950) and on the other the large debt we owe to his fascination with his "new found friends" going right back to his student days at Rutgers University (Waksman, 1954).

David Hopwood
david.hopwood@bbsrc.ac.uk

References

Bentley, S.B. *et al*. 2002. Complete genome sequence of the model actinomycete *Streptomyces coelicolor* A3(2). Nature. 417: 141-147.

Hopwood, D.A. 1999. Forty years of genetics with *Streptomyces*: from *in vivo* through *in vitro* to *in silico.* Microbiology. 145:2183-2202.

Waksman, S.A. 1950. The Actinomycetes. Baltimore, Williams and Wilkins.

Waksman, S.A. 1954. My Life with the Microbes. London, Robert Hale.

From: *Genomics of GC-Rich Gram-Positive Bacteria*

1

Genomes and Evolution

Antoine Danchin

Genomics today holds the fashionable position that molecular genetics used to occupy. Rather than work with individual genes, we now prefer to consider several genes at a time, and try to get an integrated picture of the cell or of the organism as a whole. In the present series of books, focusing on Gram positive organisms, we try to follow this state of integration by considering many facets of large-scale functional investigation in a variety of bacteria, with the explicit goal to find, in the end, common rules in their dynamic and structural organization, rules that would aid the understanding of common properties and essential differences corresponding to elusive functions, or of still unknown bacterial biotopes.

It seems indispensible, in a book devoted to genomes, to first stress the scientific reasons for genome sequencing (and, at the time of this introductory article, more than 600 public genome programs are under way). Scientific reasons have not always prevailed and we hope to provide here a strong conceptual justification to the sceptics who will then be enticed to know more by reading the articles presented in this series.

Naturally, the beginning of the Human Genome Programme is the case in point. It is important to note that this project was born of a political initiative. Begun in the mid-nineties as a full blown programme (and certainly much earlier, if we trace its origins) it very soon became inseparable from the commercial issues which surrounded it. This is so strong that, on March 14th 2000, Tony Blair and Bill Clinton published a joint declaration in which they "applaud the decision by scientists working on the Human Genome Project to release raw fundamental information about the human DNA sequence and its variants rapidly into the public domain". The declaration ended with an enigmatic phrase in which Blair and Clinton "commend other scientists around the world to adopt this policy" of rapid publication. It goes without saying that it is unusual for heads of state to intervene in scientists' decisions to publish. Incongruous as this is, the declaration reminded us that the Human Genome Project is based on a political initiative, not a scientific one (Kelly, 1990; Sinsheimer, 1990; Cook-Deegan, 1995). Is this the case with microbial genome projects? I hope that the chapters presented in this series will convince the reader that there is much more, and that genomics is in fact a revolution in the way we consider life.

This very political backgound contained an important technological insight. Genome studies are inseparable from the technique which led to DNA sequencing. Furthermore, genome sequencing has added a new dimension to the usual experimental palette we use to study life: in addition to work *in vivo* and *in vitro*, we have to work "*in silico*", to use computers (Danchin *et al.*, 1991). The enormous progress we continuously witness in genome sequencing would not have been possible without developments in computer memory and calculating speed, in parallel with improvement in sequencing techniques. As early as 1978, it had become clear that computer support was necessary to allow the scientific community to build the sequences into a continuous text which they could then interpret. A study undertaken by the Rockefeller University and the European Molecular Biology Laboratory (EMBL) at Heidelberg led to the idea of the creation of a databank for gene sequences. In fact it had even been proposed that the programme for the creation of EMBL should be the sequencing of the *Escherichia coli* genome. It became clear very early on that the possession of this information was of vital importance, with political implications. Frequent discussions, often heated, took place between Europe and the USA, to decide where these databanks would be, and how they would be structured. Who would be responsible for sequence quality – its producer or the database? Who would produce the annotations? This is clearly no small matter – a wrong annotation is tantamount to disinformation. And it is unfortunately now clear that major annotation errors are spreading via data banks through the entire scientific community. Two banks were established, in competition but also in touch with each other – one at Heidelberg, the other, the first GenBank, at one of the DOE's laboratories, the Los Alamos National Laboratory (LANL). This was a somewhat unstable situation, but the informal association between GenBank and its European and Japanese counterparts, which had existed since 1990 and which later became official, brought stability. On the European side was the EMBL data library, first at Heidelberg, then at its outstation at Hinxton Hall, south of Cambridge, the European Bioinformatics Institute (EBI), and on the Japanese side the DNA Data Bank of Japan (DDBJ) at the National Institute of Genetics (NIG) at Mishima. Effectively, there is now one single International Nucleotide Sequence Database (INSD), with three entry points at the NCBI, the EBI and the NIG.

After the first meeting on the sequencing of micro-organisms organised by David Galas at The Institute for Genomic Research, it clearly appeared that one needed a powerful computer infrastructure to develop genome programmes. David Hopwood, present at the meeting, naturally defended the idea of sequencing a *Streptomyces* genome. But at the time this was premature, in particular because it was then really difficult to sequence G+C-rich DNA. Several other important technical issues had also to be solved, and it is the "shotgun" technique, advocated by Craig Venter which demonstrated that it was possible, in a short time, to entirely sequence bacterial genomes, at least those which were neither A+T nor G+C-rich (for different reasons: the AT-rich ones lead to a large proportion of gaps because their DNA is toxic in *Escherichia coli*, while for GC-rich DNA sequencing numerous artifacts caused problems in the early days; Frangeul *et al.*, 1999). Discussions were therefore mostly about the techniques needed for the achievement of the projects, not the reasons to undertake them.

The Human Genome Programme had been explicitly conceived as a gigantic enterprise meant to construct a technological empire in the context of competition between the USA and Asia, Japan in particular. In this context, it is easy to understand the scepticism of the majority of scientists thoughout the world, althouth this is of course a mistake of scientific judgement due to the scarcity of genome programmes with a purely scientific aim. Indeed, why sequence genomes? What we shall see in this book is that not only is it a reasonable entreprise, but it is also a revolutionary turning point in our understanding of Life. Comparing genomes with one another shows in an illuminating way that the genetic programme, which specifies the RNA and the proteins, also contains an architectural programme which specifies the organisation of the cell (and of the organism). In the same way as Molecular Biology was a conceptual revolution, genomics is an even deeper revolution (why this unfortunately fashionable oxymoron "post-genomics"? "Genomics" is enough to summarize the integrative science which analyses the nature of living organisms from their genome sequence...). The history of genomes is the history of Life itself since no organism can be reduced to a single gene, nor summarized as a single family, be it expressed as a control for the cell cycle or for the secretion machinery.

In a nutshell, Life possesses a central property of self-consistency, and experience shows that this is reflected as an integration which is directly visible in the nature of the underlying genetic programme that drives the development of all cells or organisms. The various chapters of these books on Gram positive bacteria will amply demonstrate the integrated properties uncovered by large scale approaches.

The Alphabetic Metaphor

The central concept often named (wrongly, because there is no religion here!) the Central Dogma of Molecular Biology, is that 'information transfer', always associated with two processes essential for life, metabolism and compartimentalization, rests on two laws. The first one is *complementarity*, which to a sequence of primordial motifs associates a complementary sequence of motifs (chain of letters) that results in exact and complete specification of the former, and the second law is *coding*, which uses a cipher allowing the rewriting of the first chain symbolized by a four letter alphabet into a second chain symbolized by a twenty letter alphabet.

Molecular biology used to rely on the alphabetic description of the genes. It is now further exploring the metaphor, in which the genome is understood as providing the complete text of the recipe allowing the construction, the development, the survival and the evolution of any living organism. Let us note here that this would be a deep mistake ("*this is not a pipe*", painted the surrealist Magritte) to mix up the book of recipe with the meal, as one does when one tries to make the public believe that the knowledge of the Human Genome sequence will allow one to cure all diseases!

Knowing the genome texts, nevertheless, is truly revolutionary because this opens up the possibility of understanding the nature of life and of its evolution through the exploration at a symbolic level (in the same way as the letters which compose the recipe are symbolic with respect to its realisation, but have a very precise link with it). What this metaphor, which comprizes an alphabetic text and a coding process, accounts for is the possibility of a true creation (i.e. sudden appearance of an entity which cannot be predicted from what was before, but only accounted for in the chain of evolution *a posteriori*), as repeatedly witnessed when analyzing the evolution of species. Comparing genomes allows the identification of those places where creation operates: the Darwinian trio Variation / Selection / Amplification makes material systems evolve, evolution *creates* new functions, which, in order to come to being, *capture* preexisting structures — hence the "tinkering" features stressed by François Jacob. And hence the impossibility of predicting a function from the structure alone (Danchin, 1999). Naturally, because any creation is unpredictable, one will only be able to explain *a posteriori* how this or that "capture" allowed the development of that particular function. It is there that we begin to construct the means to explore, from what we know of the past, what might be conditions for new creation to come, as well as the places where it will be possible (this is a central question, for example, when one wishes to understand how and where new diseases can emerge).

The alphabetic metaphor of the genetic programme is so appropriate that one can represent the cell and its programme as a computer and its programme. That particular computer would have the particularity, in the programme it deciphers and modifies, to provide the appropriate instructions not only for its own duplication, but for the duplication of the machine itself. The most elementary machine of this type, as proposed by Alan Turing, is made of a read/write head, of a mobile "band" which passes through the head and which carries only a linear sequence of symbols, and of a mechanical device which allows the band to go forward, go backward, stop, or read the symbols in the band, as a function of the previous readings by the head. This machine is essentially defined by the fact that it allows formal separation between the *machine* proper (the read/write head and the mechanics needed to make the band move), the *data* which set the conditions under which the programme is executed, and the *programme* itself. With this view, data and programme are in fact playing the same role: they are sequences of states which are scanned by the read/write head during the functioning of the machine.

The major conceptual question asked by this metaphor is whether one is allowed to separate the programme from the machine. As I argued elsewhere (Danchin, 1998), the experiments which are at the root of the molecular techniques of biotechnology provide us with a first indication that this separation is effective. It is demonstrated in an illuminating way in these experiments of cell cloning that have been so fashionable since 1997. Thus, the metaphor raises the question of the duality that exists between the concreteness of material systems such as cells, and the symbolic nature of the systems of Number Theory (arithmetics). This led Turing for example to invent the machine we just described. For a living organism the genetic programme leads to the real construction of the machine: in this case, everything goes as if one could smoke Magritte's pipe. This indicates that we have more to understand in a genome than what we understand when we simply analyze the genes transcription and translation. We need to identify collective, global properties. Hence the need for large-scale approaches such as transcriptome and proteome analyses.

The logic needed for an effective complementarity between matter and its symbolic representation has long been discussed by von Neumann in the mid-sixties (von Neumann, 1958). It requires some "jump" from matter to abstractness. If the programme is a model for a universal construction rule that plans the duplication of both its own description and that of the machine, then the duplication of the programme together with that of the machine becomes logically possible. But this requires that the machine and its symbolic representation (programme in the programme) be somehow separated. Thus there is a need, in addition to the genetic programme, for the existence, somewhere, of an *image of the machine.* The implication, quite strong, of this line of reasoning, is that there must be, associated with the programme itself, a representation of its environnement (i.e. of the cell, and of the world where it lives). This is precisely what is allowed for by the fact that the programme has a physical support, the DNA molecule. We have indeed observed, from the analysis of the bacterial genomes text, that there is a definite correlation between the order of the genes in the genome, and the architecture of the cell (Danchin *et al.*, 2000).

Bacterial Genomes and Their Architecture

It is frequently heard that the distribution of the genes in genomes is more or less random, that genomes are "fluid" entities (Stibitz and Yang, 1997). In fact, a superficial observation of the position of orthologous genes in various genomes gives the impression that they can be located anywhere, and certainly not always at the same position in different genomes (this is difficult to assess since one would need for such an investigation to locate genes with respect to appropriate reference points, but are there "centromeres" in bacteria, for example?). However, it is also clear that genes have a real tendency to form clusters: genes coding for related functions, such as those coding for the subunits of an enzyme for example, are most often located side by side, in the same transcription unit. This was the very basis of the concept of *operon*. This grouping is certainly not random. This may even sometimes give clues for the identification of an unknown function (Nitschké *et al.*, 1998). Here is an example that is rarely discussed despite its reference to the paradigm of transcription expression and control. The full meaning of the lactose operon, formed of three genes, *lacZYA*, respectively coding for beta-galactosidase, lactose permease, and lactose transacetylase long remained elusive: what is the function of transacetylase? This is particularly puzzling, knowing that the affinity of this enzyme for lactose is so poor (K_M of about 1 M, Zabin and Fowler, 1984). A logical analysis of the operon shows that if the permease is very efficient (and it is!), then the concentration of lactose in the cell may reach such a high level that it will become toxic, be it only because lactose influx will increase osmolarity to an unbearable point. One needs therefore to think of a mechanism of excretion (Liu *et al.*, 1999). But one also needs to understand that the cell has to avoid a futile cycle, so that the excreted molecule cannot be the one which enters the cell: the formation of acetyl-lactose is thus likely to be the much needed security valve!

What do we see when we pay heed to the situation? In bacteria, in general, the chromosome is replicated from a unique replication origin. This origin can thus be used as a reference point. Furthermore, in these organisms, many genes are co-transcribed in the form of operons. When one takes the origin as the reference, and the existence of operons as markers of the gene distribution, then the fluid picture of the genomes changes considerably. Indeed, one first remarks that there are almost always more genes transcribed in the direction of the movement of the replication fork than in the opposite direction (this varies much from genome to genome: organisms such as mycoplasma or A+T-rich Gram positive bacteria have a very strong transcription bias, whereas this is much less true in *Escherichia coli* for example). Second, one remarks that, quite often, the nature of the base composition in each of the chromosome strands is very different: it is G+T-rich in the direct strand and A+C-rich in the complementary strand (this composition bias, when it exists, is the same for all bacteria) (Frank and Lobry, 1999; Rocha *et al.*, 1999). If the mutation rate had time to equilibrate (which would be the case if genes really moved frequently; Rocha and Danchin, 2001) then, as shown by Sueoka, one should have within each strand an equal number of A and T, and of G and C (the second Chargaff's law), as is observed between the strands (Sueoka, 1995).

This demonstrates that genes do not move that frequently, or, more precisely that most gene moves are incompatible with the long term survival of the species. What we see as a move represents only what remains, keeping compatibility with long term survival, keeping mainly a privileged orientation (and maybe privileged distances) with respect to the origin of replication. As a consequence, if one looks carefully, one observes in bacteria a strong constraint in the distribution of the genes with respect to the origin and terminus of replication. Furthermore, the analysis of the codon usage bias indicates that there exist many additional constraints. Indeed, we have shown in 1991 that a significant part of the *E. coli* genome corresponded to genes with an unusual codon usage bias. These genes were most probably involved in horizontal gene transfers, and they were many, forming one sixth of the genome in *E. coli* K12 (Médigue *et al.*, 1991). Since then, this observation has been made over and over again in almost every bacteria, sometimes for a considerable proportion of the genome, as is the case for example in some *Bacillus cereus* species which may thus be considered as large reservoirs for genes of unknown functions (not visible in the phenotype of the colonies on a plate, nor in the classical tests for systematic identification; Carlson and Kolsto, 1994).

At first sight, these genes are distributed randomly. One remarks however that they are more numerous near the replication terminus, which is therefore a hot spot for input of foreign DNA in a genome (the mechanics of recombination needed to resolve the knotted stucture which forms when replication terminates accounts for this phenomenon; Corre *et al.*, 1997). However laterally transferred genes appear to be present almost anywhere. There does not yet exist a careful study of their distribution, but one must remark that these “foreign” genes are not randomly placed. As a matter of fact, when one goes from one strain to another one (for example comparing *E. coli* K12 and *E. coli* O157H7), one remarks that the foreign genes differ, but that they are placed at the same position in the chromosome, often near tRNA genes (Hou, 1999; Karch *et al.*, 1999; Blum-Oehler *et al.*, 2000; Hacker and Kaper, 2000; Al-Hasani *et al.*, 2001). This, again, is compatible with the notion of a strong architectural constraint in the gene distribution. It is as if one witnessed a rigid organisation of the chromosome, interspersed with a certain number of hot spots where foreign genes can insert.

Careful study of the codon usage bias further points to the existence of strong constraints in the gene order. Indeed, when one analyzes the genes within an operon, one usually observes that the corresponding codon usage bias is the same in its different genes. However, and this is remarkable, when one compares two organisms where genes are clustered into an operon in one of them while the corresponding genes are dispersed in the other, one finds that the codon usage bias is preserved, even when the the genes are no longer clustered but located distantly (Rocha *et al.*, 2000a). How can we account for this observation? The usage of a particular codon assumes that a particular transfer RNA is used. It is therefore easy to see that repeating the same codon in a gene tends to maintain the cognate tRNA local concentration at a high level (in the vicinity of the ribosome that translates the messenger RNA). A large bias in the codon usage implies therefore a local enrichment in some tRNAs, while in the same local environment the corresponding ribosome is depleted in other tRNAs. A polycistronic operon is usually translated by the same ribosomes (Petersen *et al.*, 1978). One understands therefore that the genes it is made of have the same codon usage bias (otherwise this would introduce a pause in the translation process, needed for the appropriate tRNA to become available: a fact probably used for folding proteins; Thanaraj and Argos, 1996). But the fact that genes distant in the chromosome share the same bias, strongly suggests that the corresponding mRNAs are sitting next to each other, and are translated from the same or very closely packed ribosomes… This indicates that there is a relation between the position of some ribosomes in the cell, and the position of some genes in the chromosome. There is a map of the cell in the chromosome.

A complementary observation strengthens this hypothesis considerably. The analysis of the codon usage bias in orthologous genes in organisms as different as *E. coli* and *B. subtilis*, shows a similar deviation of the bias with respect to its average value (different in both bacteria) for genes with orthologous functions, indicative of the existence of a common selection pressure in these quite different organisms (Danchin *et al.*, 2000). Apart from an architectural selection pressure what might be the cause of such a strong correlation? Naturally, for selection pressure to operate, one needs a constant and repeated action from the environment, and this can only happen for genes that are highly expressed, in conditions where the progeny of the cell is large (as happens during the exponential growth phase), so as to ensure amplification of the tiny differences in selection. The relevant genes are therefore those which are expressed at high level during exponential growth. Many of these genes are those which constitute the core of life, being responsible for the transcription and the translation of the genetic programme or for the general management of energy. Indeed, these genes are usually clustered into well recognizable large operons. The first step in the analysis of all newly obtained genome sequence should therefore consist in the fine analysis of the conservation of the genes neighborhoods. It is highly probable that, in the future, *specific* laws of this organisation will be uncovered. This becomes therefore a well defined goal for genomics. The investigation of the role, position and architecture of pathogenicity islands (Dobrindt and Reidl, 2000; Hacker and Kaper, 2000) is a first step in this exploration.

Evolution of Genomes

If one understands which genes are submitted to this architectural selection pressure (one must add the genes of the core intermediary metabolism) it is necessary to attempt to identify the nature of the selection pressure that maintains them organized with respect to each other. To this aim we need to think that this probably already happened at the origin of the first cells, and persisted in present day organisms despite the very large time which elapsed since then.

Many scientists asked questions about the origin of life, and some understood that the hypothesis of a prebiotic soup (the very soup that Pasteur demonstrated to be incapable of spontaneous generation!) would act as a poisoned broth unable to allow life to be born (Danchin, 1989). It was necessary therefore to propose other scenarios. Steven Benner used to state that "*arguments that attempt to extrapolate from modern biochemistry back to the origin of life are futile*" (Benner *et al.*, 1987). However Granick first (Granick, 1957), followed by Wächtershäuser who advocated "*a reconstruction of precursor pathways by retrodiction from extant pathways*" (Wächtershäuser, 1988), have shown that the most plausible one was to considerer first extant metabolic processes (one must construct the machine and maintain it) and subsequently to make the hypothesis that, despite the many years of evolution, some characters reminiscent of the origin were still present today. Briefly, the most plausible is to think that a selective process, which kept mainly charged molecules at the surface of solids (hence the ubiquitous character of the phosphate and carboxylate groups in anabolic metabolites) was the initial driving force. Both authors, furthermore, also thought that, as electron transfers are central to the management of energy, iron and sulfur are at the core of the primaeval metabolism (Wächtershäuser, 1992), sulfur being also a remarkable element which allows the transfer of all kinds of chemical groups (Danchin, 1989; Sekowska *et al.*, 2000).

The current working hypothesis is that metabolism organized itself around nuclei of iron-sulfur clusters, and more generally around sulfur metabolism, at the time when the first nucleotides were invented, together with the first coenzymes, then the first membranes and the first nucleic acids. It is therefore necessary to explore first within the genomes (and in particular within those of prototrophic organisms, because they are unlikely to result from the degradation of richer ancestral genomes), those genes which make these essential metabolic pathways. In *E. coli*, *in silico* genome analysis shows that the genes of sulfur metabolism are not randomly distributed but form islands (Rocha *et al.*, 2000b). One does not know yet the homologous genes in other bacteria, but with the little information already in our possession it appears clearly that, here too, these genes form islands.

Where does the corresponding selective constraints originate from? A simple observation allows one to propose a conjecture to account for it. Most genomes contain a very large proportion of unknown genes, or of genes similar to genes of unknown function. This observation, made simultaneously by the consortium sequencing the yeast chromosome III and the scientists sequencing the first long fragment of the genome of *B. subtilis*, was quite unexpected (presented in 1991 at the European Union meeting organized in Elounda, Greece). Indeed, the existence of many mutant types with the identification of the cognate genes let everybody assume at this time that one knew all the major protein categories with their associated functions. What are these genes? Some thought initially that they corresponded to unknown regulatory systems, to unexpected cell compartments or to new functions, such as clocks or timers. However it appears that as we progress in uncovering new functions, many among these genes code for ... enzymes! How then did they escape attention? The explanation, perhaps, is simple: the biochemical study of enzymes is readily available in the laboratory using pure products with straightforward ways to measure their concentration and availability. Two classes of molecules resist these experimental constraints, the gases and the radicals. Indeed many of these "new" genes code for functions involving gases and radicals. This leads to an important question: gases diffuse fast, and many are quite reactive: putting H_2S and O_2 together (with an appropriate catalyst) makes them react violently.

How does the cell cope with this situation? The answer provides us with the missing link between the architecture of the cell and the distribution of the genes in the genome: what led to the organized clustering of genes in genomes is the selection pressure due to the reactivity and diffusibility of gases and radicals. This selection pressure had a major impact, there must exist a link between the chromosome, and the location of the product of translation in the cell. While this is easily visualized in procaryotes (because transcription is not separated from translation, allowing an organized network of mRNAs to form in relation with the construction of protein complexes), this is much more difficult to visualize when there is a nucleus. One is thus led to ask whether the role of introns is not, at least in part, to allow the formation of such a functional relation (if splicing is performed not in a more or less random way in the nucleus, but at precise loci associated with nuclear pores well placed at the nuclear envelope). But this is another story: we are concerned here only with bacteria.

Conclusion

As just discussed, we now face a new picture of the organisation of bacterial genomes. Of course this picture should be completed by studies in the domain of other bacteria such as the Archea, or the bacteria filled with complex membrane structures such as the cyanobacteria. The situation of G+C rich Gram positives is particularly interesting since it comprises bacteria growing very slowly such as *Mycobacterium tuberculosis* or *M. leprae*, as well as bacteria which differentiate into complex structures such as the *Streptomyces* sp. At the date when this introduction was written Nikos Kyrpides (http://wit.integratedgenomics.com/GOLD/) identifies 594 genome sequecing programmes (among which 93 are already complete and published). We are just beginning to investigate the surface of the information they contain. The articles in this series will present focused views on some of the topics mentioned above. Let us hope that this will be used by our colleagues as a starting point for exploring further the deep meaning of genome sequences and above all couple it to biological meaning.

References

Al-Hasani, K., Adler, B., Rajakumar, K., and Sakellaris, H. 2001. Distribution and structural variation of the she pathogenicity island in enteric bacterial pathogens. J. Med. Microbiol. 50: 780-786.

Benner, S.A., Allemann, R.K., Ellington, A.D., Ge, L., Glasfeld, A., Leanz, G.F., Krauch, T., MacPherson, L.J., Moroney, S., Piccirilli, J.A., and et al. 1987. Natural selection, protein engineering, and the last riboorganism: rational model building in biochemistry. Cold Spring Harb. Symp. Quant. Biol. 52: 53-63.

Blum-Oehler, G., Dobrindt, U., Janke, B., Nagy, G., Piechaczek, K., and Hacker, J. 2000. Pathogenicity islands of uropathogenic E. coli and evolution of virulence. Adv. Exp. Med. Biol. 485: 25-32.

Carlson, C.R., and Kolsto, A.B. 1994. A small (2.4 Mb) Bacillus cereus chromosome corresponds to a conserved region of a larger (5.3 Mb) Bacillus cereus chromosome. Mol. Microbiol. 13: 161-169.

Cook-Deegan, R.M. 1995. Origins of the Human Genome Project. WWW Article: http: //www.fplc.edu/risk/vol5/spring/cookdeeg.htm.

Corre, J., Cornet, F., Patte, J., and Louarn, J.M. 1997. Unraveling a region-specific hyper-recombination phenomenon: genetic control and modalities of terminal recombination in Escherichia coli. Genetics 147: 979-989.

Danchin, A. 1989. Homeotopic transformation and the origin of translation.

Prog. Biophys. Mol. Biol. 54: 81-86.

Danchin, A. 1998. La Barque de Delphes. Odile Jacob, Paris. (*Translation:* The Delphic Boat. Harvard University Press, 2003.)

Danchin, A. 1999. From protein sequence to function. Curr. Opin. Struct. Biol. 9: 363-367.

Danchin, A., Guerdoux-Jamet, P., Moszer, I., and Nitschké, P. 2000. Mapping the bacterial cell architecture into the chromosome. Philos. Trans. R. Soc. Lond. B Biol. Sci. 355: 179-190.

Danchin, A., Médigue, C., Gascuel, O., Soldano, H., and Hénaut, A. 1991. From data banks to data bases. Res. Microbiol. 142: 913-916.

Dobrindt, U., and Reidl, J. 2000. Pathogenicity islands and phage conversion: evolutionary aspects of bacterial pathogenesis. Int. J. Med. Microbiol. 290: 519-527.

Frangeul, L., Nelson, K.E., Buchrieser, C., Danchin, A., Glaser, P., and Kunst, F. 1999. Cloning and assembly strategies in microbial genome projects. Microbiology 145: 2625-2634.

Frank, A.C., and Lobry, J.R. 1999. Asymmetric substitution patterns: a review of possible underlying mutational or selective mechanisms. Gene 238: 65-77.

Granick, S. 1957. Speculations on the origin and evolution of photosynthesis. Annals New York Acad. Sci. 69: 292-308.

Hacker, J., and Kaper, J.B. 2000. Pathogenicity islands and the evolution of microbes. Annu. Rev. Microbiol. 54: 641-679.

Hou, Y.M. 1999. Transfer RNAs and pathogenicity islands. Trends Biochem. Sci. 24: 295-298.

Karch, H., Schubert, S., Zhang, D., Zhang, W., Schmidt, H., Olschlager, T., and Hacker, J. 1999. A genomic island, termed high-pathogenicity island, is present in certain non-O157 Shiga toxin-producing Escherichia coli clonal lineages. Infect. Immun. 67: 5994-6001.

Kelly, E.H. 1990. The Human Genome Initiative: a different type of research. Faseb J. 4: 1423-1424.

Liu, J.Y., Miller, P.F., Willard, J., and Olson, E.R. 1999. Functional and biochemical characterization of Escherichia coli sugar efflux transporters. J. Biol. Chem. 274: 22977-22984.

Médigue, C., Rouxel, T., Vigier, P., Hénaut, A., and Danchin, A. 1991. Evidence for horizontal gene transfer in Escherichia coli speciation. J Mol Biol 222: 851-856.

Nitschké, P., Guerdoux-Jamet, P., Chiapello, H., Faroux, G., Hénaut, C., Hénaut, A., and Danchin, A. 1998. Indigo: a World-Wide-Web review of genomes and gene functions. FEMS Microbiol. Rev. 22: 207-227.

Petersen, H.U., Joseph, E., Ullmann, A., and Danchin, A. 1978. Formylation of initiator tRNA methionine in procaryotic protein synthesis: in vivo polarity in lactose operon expression. J. Bacteriol. 135: 453-459.

Rocha, E.P., and Danchin, A. 2001. Ongoing evolution of strand composition in bacterial genomes. Mol. Biol. Evol. 18: 1789-1799.

Rocha, E.P., Danchin, A., and Viari, A. 1999. Universal replication biases in bacteria. Mol Microbiol 32: 11-16.

Rocha, E.P., Guerdoux-Jamet, P., Moszer, I., Viari, A., and Danchin, A. 2000a. Implication of gene distribution in the bacterial chromosome for the bacterial cell factory. J. Biotechnol. 78: 209-219.

Rocha, E.P., Sekowska, A., and Danchin, A. 2000b. Sulphur islands in the Escherichia coli genome: markers of the cell's architecture? FEBS Lett. 476: 8-11.

Sekowska, A., Kung, H.F., and Danchin, A. 2000. Sulfur metabolism in Escherichia coli and related bacteria: facts and fiction. J Mol Microbiol Biotechnol 2: 145-177.

Sinsheimer, R.L. 1990. Human genome initiative. Science 249: 1359.

Stibitz, S., and Yang, M.S. 1997. Genomic fluidity of Bordetella pertussis assessed by a new method for chromosomal mapping. J. Bacteriol. 179: 5820-5826.

Sueoka, N. 1995. Intrastrand parity rules of DNA base composition and usage biases of synonymous codons. J. Mol. Evol. 40: 318-325.

Thanaraj, T.A., and Argos, P. 1996. Ribosome-mediated translational pause and protein domain organization. Protein Sci. 5: 1594-1612.

von Neumann, J. 1958 (reed. 1979). The Computer and the Brain. Yale University Press, New Haven.

Wächtershäuser, G. 1988. Before enzymes and templates: theory of surface metabolism. Microbiol Rev 52: 452-484.

Wächtershäuser, G. 1992. Groundworks for an evolutionary biochemistry: the iron-sulphur world. Prog. Biophys. Mol. Biol. 58: 85-201.

Zabin, I., and Fowler, A.V. 1984. Purification of thiogalactoside transacetylase by affinity chromatography. Anal. Biochem. 136: 493-496.

From: *Genomics of GC-Rich Gram-Positive Bacteria*

2

Genetic Organization of *Rhodococcus*

Leonid A. Kulakov and Michael J. Larkin

Abstract

The genus *Rhodococcus* contains diverse bacterial species inhabiting a variety of environmental niches ranging from polluted soils to plants and animals. It is generally acknowledged that rhodococci play a key role in nature by recycling complex organic compounds including a wide range of aromatic hydrocarbons and most persistent xenobiotics. Although our knowledge of the genetic organization of *Rhodococcus* is not as complete as for some other groups of bacteria, there were many significant advances in recent years. Reviewed here is the current knowledge of rhodococci genome organization. In particular the roles of linear and circular plasmids, mobile genetic elements and genomic rearrangements are considered. Further genetic and molecular biology studies of *Rhodococcus* species will be essential for the successful exploitation of these bacteria.

Introduction

The genus *Rhodococcus* encompasses diverse bacterial species. The members of this group are found in a variety of environmental niches such as pristine and polluted soils, sea basins, plants and animals (Finnerty, 1992). Rhodococci are aerobic, Gram-positive to Gram-variable, non-motile, catalase-positive actinomycetes. Classification of rhodococci is based upon cell wall structure which is characterized by the presence of mycolic acids (Goodfellow, 1989). For many years *Rhodococcus* species have been difficult to identify due to a lack of suitable phenotypic tests (Goodfellow *et al*., 1998). More recently 16SrRNA sequences for the number of *Rhodococcus* species were obtained. Analysis of the sequencing data established the close phylogenetic relationship between the genera *Rhodococcus*, *Corynebacterium*, *Nocardia* and *Mycobacterium* (Rainey *et al*., 1995). Sequencing analysis also confirmed the view that *Rhodococcus* species should be assigned to two groups each of which merited the status of genera (Goodfellow, 1989). Additionally, as the result of this approach the taxonomic position of many actinomycetes have been changed, i.e. some strains of *Nocardia* were reclassified as *Rhodococcus* (Yoon *et al*., 1997) and also a number of former *Rhodococcus* species were placed to other genera (Wang *et al*., 1995).

The importance of *Rhodococcus* and the related group of *Nocardia* was first recognized due to the presence of opportunistic pathogenic strains and especially for their veterinary significance (Finnerty, 1992). It is widely recognized now that species of the genus *Rhodococcus* play a key role in nature by recycling complex organic compounds including a wide range of aromatic hydrocarbons (Finnerty, 1992; Larkin *et al*., 1998). The ability of rhodococci to degrade a wide range of xenobiotics and their persistence in various environmental niches makes them ideal candidates for the bioremediation of numerous contaminated sites. A number of important biotransformations can be catalysed by *Rhodococcus* species (Warhurst and Fewson, 1994) and some strains also produce surfactants which are considered as environmentally friendly (Lang and Philp, 1998).

Despite the acknowledged significance of the genus in the environment and its growing significance in industrial processes the genetic organization of rhodococci is still poorly understood. A number of obstacles that hinder the progress of genetic analysis of *Rhodococcus* were pointed out by Larkin *et al*., (1998). Amongst these is the immense genetic diversity of the group, rendering it difficult to select a type strain for studies, and the remarkable genomic instability of many *Rhodococcus* characteristics.

To date there has been no complete genome sequence freely published for *Rhodococcus*. However, the economic significance of these bacteria is exemplified in the completion of the genome sequence of a commercial strain, IG24 employed by the Merck Corporation in developing a biotransformation route for the production of the anti HIV drug Crixivan®. The sequencing of this strain has been carried out by Integrated Genomics Inc (Chicago. IL) and the total genome size is reported to be 5487 kb with 6103 ORFs. (http://www.integratedgenomics.com/contact.html). However, the sequence is not openly available and may in any case prove to be of limited use due to the wide variation in genome size noted for these bacteria. In contrast, work has begun in 2001 to determine the complete genome sequence of *Rhodococcus* strain RHA1. This is being carried out by Davies and coworkers at the University of British Columbia, Vancouver as part of the CanadaGenome initiative (Microbial Envirogenomics: Micro-organisms and their Interaction with the Environment http://www.genomecanada.ca/index.htm).

Of particular interest is the distribution of large linear plasmids amongst the rhodococci. This is slowly being revealed by various workers as a very common feature of these bacteria and their full diversity and significance has yet to be elucidated. *Rhodococcus* strain RHA1 was originally isolated as a polychlorinated biphenyl (PCB) degrader and has been shown to possess large linear plasmids, including pRHL1 (1,100 kb), pRHL2 (450 kb), and pRHL3 (330 kb) (Shimizu *et al.*, 2001). Thus, the determination of the complete sequence of this particular strain will provide considerable insight into the organization of genes located in chromosome and linear plasmids. Here we provide an overview of what has been determined to date concerning the genetic organization and genome structure of the rhodococci.

Chromosomes and Plasmids

Research on *Rhodococcus* genetics dates back to the 1960's and at that time the first linkage maps of chromosomal markers for *R. erythropolis* (Adams, 1964) were constructed. However, in comparison with *Streptomyces*, where the complete sequence of the genome is reported (Ferretti *et al.*, 2001), knowledge of the genetic organization of *Rhodococcus* species still remains incomplete and fragmented.

A linear chromosome of about 1 Mb has been reported for *R. fascians* (Crespi *et al.*, 1992). There is also some evidence that the four mega-base chromosome of *Rhodococcus fascians* D188 may be linear as well (Crespi *et al.*, 1992). However, other analysis of several virulent and avirulent strains of *R. fascians* did not detect any linear replicons in the megabase range, (Pisabarro *et al.*, 1998). The genome size estimated by the same authors varied for different strains and was in a range of 5.6 - 8.0 Mb (Pisabarro *et al.*, 1998). These reported differences in the genome size may be related to the presence of plasmids and instability of the *Rhodococcus* genome (see below). There are few reports on the analysis of the chromosome size in other *Rhodococcus* species.

Generally, information about the organization of the genomes of actinomycetes is limited to a few genera. *Streptomyces* species and *Saccharopolyspora erythraea* appear to contain a single linear 8 Mb chromosome. In contrast, bacteria of the genera *Corynebacterium, Mycobacterium* and *Rhodococcus* appear to possess smaller (3 Mb-6.5 Mb) single circular chromosomes in addition to linear plasmids. It has been suggested that large genome sizes and chromosome linearity may be limited to the mycelium forming actinomycetes (Redenbach *et al.*, 2000).

Unsurprisingly, plasmids are common in *Rhodococcus* species. Indeed the majority of *Rhodococcus* strains analyzed to date harbor plasmids, often with many different types of plasmid present in the same cells. Linear plasmids similar to those found in many *Streptomyces* species (e. g. Hayakawa *et al.*, 1979; Meinhardt *et al.*, 1997) are also abundant in *Rhodococcus* strains (Crespi *et al.*, 1992; Kalkus *et al.*, 1990; 1993). A common feature of the linear replicons of many *Streptomyces* is the presence of inverted repeats with proteins covalently bound to the 5' end of each DNA strand (Hinnebusch and Tilly, 1993) The telomers of several *R. opacus* linear plasmids with sizes of around 200 kbp have been analyzed in detail. Plasmid pHG207 (225 kbp) possessed all of the features typical for linear plasmids of *Streptomyces*, however its terminal inverted repeats (TIR) of 583/560 bp are relatively short (Kalkus *et al.*, 1993). The plasmids pHG201, pHG204 and pHG205 were also found to carry proteins attached to their ends. However, the TIRs of these plasmids were significantly shorter than those of the large linear replicons of *Streptomyces* and pHG201 was shown to have no defined TIR at all. In that case the homology of the terminal 34/32 bases was only 65% (Kalkus *et al.*, 1998). Various genes for biphenol and polychlorinated biphenol (PCB) degradation have been located on the linear plasmids pRHL1 (1,100 kb) and pRHL2 (450 kb) in *Rhodococcus* sp. RHA1 (Shimizu *et al.*, 2001). A physical map was constructed for pRHL2 and the termini of this plasmid were analyzed. It was shown that both termini were linked to a protein, and that the TIRs showed limited similarity to each other (64% identity) (Shimizu *et al.*, 2001).

Various *Rhodococcus'* functions such as hydrogen autorophy in *R. opacus* (Kalkus *et al.,* 1990; 1993), isopropyl benzene metabolism in *R. erythropolis* (Dabrock *et al.,* 1992), biphenyl metabolism in *Rhodococcus sp.* HA1 (Masai *et al.,* 1997) the plant virulence/fasciation genes (*fas*) in *R. fascians* (Crespi *et al.* 1992) and others have been reported to be encoded by linear plasmids. However, more linear plasmids with unknown functions were detected in other isolates: *Rhodococcus corallinus* (Archer and Griffiths 1995), *R. rhodochrous* NCIMB13064 and a wide variety of soil isolates which degrade aromatic substrates (Kulakov and Larkin, unpublished).

Circular plasmids are also very common in *Rhodococcus* species. The most extensively characterized plasmids are either associated with pathogenesis determinants of *R. equi* (Prescott, 1991; Takai, 1997) or control the degradation of aromatic compounds in various strains (Kulakova *et al.*, 1997). Some strains have different types of plasmids. For example, *R. opacus* MR11 and MR12 have been shown to have a mixture of large linear and circular plasmids (Kalkus *et al.*, 1990; 1993). A similar variety of plasmids was detected in *R. rhodochrous* NCIMB13064 (Kulakov and Larkin, unpublished). Additionally, small cryptic plasmids have been characterized in various *Rhodococcus* strains. The complete sequences of two small circular cryptic plasmids, pKA22 and pFAJ2600, were reported (Kulakov *et al.*, 1997; De Mot *et al.,* 1997). Some of these plasmids appear to be associated with certain rearrangements of the bacterial genome. Two cryptic plasmids pKA22 (5 kb) and pKA4 (9 kb) were found in *R. rhodochrous* NCIMB13064 derivatives that had lost the ability to utilize short-chain 1-chloroalkanes but had acquired naphthalene-degrading ability. The other plasmid, pFAJ2600 (6 kb), is a stable cryptic plasmid found in *R. erythropolis* NI86/21 (De Mot *et al.*, 1997). Apart from clear similarities of the putative replication regions, there is no further sequence homology between pKA22 and pFAJ2600. Two putative stability loci identified in FAJ2600 (De Mot *et al.,* 1997) are not present in pKA22. Based on the characteristics of their putative replication genes, both plasmids were assigned to the pAL5000 family of replicons identified in members of the genera *Bifidobacterium, Brevibacterium, Corynebacterium, Mycobacterium* and *Neisseria.*

In some cases *Rhodococcus* plasmids have been shown to be transferred by conjugation to the derivatives of the parental strain (Kulakova *et al.*, 1995) as well as to the other species.

Replication of *Rhodococcus* Plasmids

A significant number of *Rhodococcus* plasmids bearing genetic determinants related to biodegradation, biotransformation, antibiotic and metal resistances, and virulence have already been characterized, but little effort has been made to study the replicons themselves. The genetic control of replication, stable maintenance, and replicon transfer is poorly understood. No determinants related to chromosomal replication were reported for *Rhodococcus*. This information, however, is very important for a rational design of cloning vectors and would also be essential to elucidate gene mobility and ultimately evolution of these bacteria.

It was shown that the putative RepA proteins of *Rhodococcus* plasmids pKA22 (Kulakov *et al*., 1997) and pFAJ2600 (De Mot *et al.,* 1997) as well as that from *Mycobacterium fortuitum* plasmid pAL5000 were related to RepA proteins of the enterobacterial ColE2-type plasmids. This similarity suggested that *Rhodococcus* encoded RepA proteins may act as plasmid specific primases, which synthesize unique RNA primers (Takechi *et al*., 1995). Studies of plasmid pAL5000 revealed that the *rep*AB genes are required for replication (Stolt and Stoker 1996a). Analysis of the RepB protein from pFAJ2600 also showed its similarity to that of pAL5000 (De Mot *et al.,* 1997). The RepB protein of pAL5000 binds specifically to two DNA sites separated from each other by 37 nucleotides and located upstream of *rep*A gene (Stolt and Stoker 1996b). The low-affinity binding site (which is immediately upstream of the conserved 15-bp element in pAL5000; see below) is probably part of the origin of replication and the high-affinity binding site presumably enables auto-regulation of *rep*B expression (Stolt and Stoker 1997). De Mot *et al.* (1997) found two putative DNA binding domains in the RepB protein of pFAJ2600 plasmid which were similar to ColE2 proteins C-terminal regions involved in binding to the origin of replication (Shinohara and Itoh 1996). This observation suggests that the *Rhodococcus* RepB DNA binding protein may act together with a primase (RepA) in a manner similar to the Rep proteins of ColE2 plasmids when initiating DNA replication (De Mot *et al.,* 1997). The existence of the ColE2-type origin of replication (*ori*) in the pAL5000-related plasmids was further confirmed by the presence of copies of a conserved 15-bp sequence immediately upstream of *rep*A (De Mot *et al.,* 1997). These sequences resemble the *ori* site of the ColE2 plasmids, where unidirectional leading strand synthesis is initiated following synthesis of a specific primer by the DNA-bound Rep protein (Hiraga *et al.,* 1994; Takechi and Itoh 1995; Takechi *et al.,* 1995). In pKA22, a similar sequence forms a part of a long direct repeat located upstream of ORF2, which is a homologue of RepA (Kulakov *et al.*, 1997). When aligned, RepA proteins from pKA22 and pFAJ2600 showed low homology, even though both are encoded by *Rhodococcus* replicons. However, ORF2 (RepA) of pKA22 is more similar (60 % identity) to RepA of pAL5000 (Kulakov *et al.*, 1997) than the RepA of pFAJ2600 which shares only 33 % identical residues with the same protein (De Mot *et al.,* 1997). A replication region was also analyzed for the 100-kbp plasmid (pSOX), encoding desulfurization, which was isolated from *Rhodococcus sp* X309 (Denis-Larose *et al.*, 1997). The putative replication (Rep) protein sequence of this plasmid was found to be related to Rep proteins of pLR7 family of plasmids characterized in *Mycobacteria* (Denis-Larose *et al.*, 1998). Analysis of these two groups of *Rhodococcus* replicons may lead to the conclusion that a number of *Rhodococcus* and *Mycobacteria* plasmids have a common origin. However, in contrast to the examples discussed above it has been shown that the *ori* region of the *R. equi* virulence associated plasmid pOTS bears no similarity with those of *Mycobacteria* or any other known plasmids. No proteins are known which show homology to the putative Rep protein found in this region (Zheng *et al.,* 1997).

Mobile Elements

Analysis of mobile elements and the development of transposition systems plays an important role in advancing the genetic analysis and understanding of the genome organization of any group of bacteria. Transposons so far have not been described for *Rhodococcus* species. However, such elements are likely to be abundant in this group of microorganisms since sequencing analysis of different regions of rhodococcal genomes often results in discoveries of new putative mobile elements. Although a limited number of insertion sequences (IS) have been identified and analyzed in rhodococci, they show significant diversity and their contribution to the acquisition of biodegradative and other phenotypes by promoting mutations and genomic rearrangements in both chromosomes and transmissible plasmids seems highly probable.

The *Bacillus subtilis sac*B gene that encodes the production of levansucrase (Steinmetz *et al.*, 1985) is widely used for the selection of insertion sequences in different bacteria, because of its lethal effect in the presence of sucrose. This approach was also successfully employed for the isolation of an IS element (tentatively named IS-*Rf*) from *R. fascians* (Jäger *et al.,* 1992, 1995). IS-*Rf* has not been fully characterized so far. It seems likely that this type of selection for IS elements may be applied to other *Rhodococcus* species (Denis-Larose *et al.*, 1998). However, most of the mobile elements of *Rhodococcus* were identified when sequencing analysis of DNA regions flanking genes of interest was undertaken. Three mobile elements discovered this way were: IS*1166* located downstream of the sulfur oxidation genes (*soxABC*) in *R. erythropolis* IGTS8 (Denome and Young 1995); IS*1164* found upstream of the structural genes for the high molecular mass-nitrile hydratase (*nhhBA*) in *R. rhodochrous* J1 (Komeda *et al.,* 1996a); and an IS element detected upstream of the hydrogenase subunit genes (*hoxFUYH*) of *R. opacus* MR11 (Grzeszik *et al.,* 1997). All these IS elements belong to the IS*256* family members of which were identified in a wide range of organisms (Mahillon and Chandler, 1998). The detection of multiple copies of these elements in some of the strains (Denome and Young, 1995; Komeda *et al.,* 1996b) suggested their ability to be effectively transposed in *Rhodococcus*. Hybridization analysis showed the presence of one or more sequences homologous to IS*1164* in members of several *Rhodococcus* species (Kobayashi *et al.*, 1997). Denis-Larose *et al.* (1997) discovered copies of IS*1166* in a number of different sulfur oxidising strains of *Rhodococcus* and IS elements of the IS*256* family are also widespread in members of the genus *Mycobacterium* (Picardeau *et al.,* 1996, 1997).

Two copies of the same insertion sequence (IS*2112*) were identified on plasmid pRTL1, encoding enzymes for the utilization of 1-chloroalkanes by *Rhodococcus rhodochrous* NCIMB13064 (Kulakov *et al.*, 1999). Another putative transposase gene was found in the same region immediately upstream from haloalkane dehalogenase gene (*dhaA*) of *R. rhodochrous* NCIMB13064 and its involvement into rearrangements of *dha*A region seems probable. IS*2112* belongs to the IS*110* family of transposable elements and, along with IS*116* (Leskiw *et al.,* 1990), IS*2112* does not contain terminal inverted repeats and is not flanked by directly repeated sequences; it forms a separate subfamily of IS elements (Kulakov *et al.*, 1999). Insertion elements of the IS*110* family are found in members of *Mycobacterium* and *Streptomyces* species. All members of this family have none or very short (<7 bp) inverted repeats and usually do not generate direct repeats in target sites (Green *et al.,* 1989; Leskiw *et al.,* 1990; Mahillon and Chandler, 1998).

Nagy *et al.* (1997b) discovered another element, IS*1415* in the DNA region upstream of the cobalamin biosynthesis genes (*cobLMK*) in *R. erythropolis* NI86/21. Three copies of this element, belonging to the IS*21* family (Haas *et al.,* 1996), have been found in strain NI86/21. A mini-transposon (Tn*5561*) was constructed by inserting the chloramphenicol resistance gene *cmrA* derived from pDA71 plasmid into IS*1415*. This construct successfully transposed following delivery on a suicide vector to *R. erythropolis* SQ1 and low specificity of the target sequences was demonstrated. Sequence analysis of insertion sites in strains NI86/21 and SQ1 revealed a 5- or 6-bp duplication of the target sequences. The apparent lack of sequence preference at the target site suggests that this system may be valuable for transposon mutagenesis in *Rhodococcus*. However, transposition from an non-replicating vector introduced by electroporation was rather inefficient (Nagy *et al.*, 1977b). A conditionally replicating delivery system has been proven successful in *Mycobacterium* (Guilhot *et al.,* 1994; Bardarov *et al.,* 1997; Pelicic *et al.*, 1997). This approach could significantly improve the efficiency of transposition mutagenesis in *Rhodococcus*.

A sequence (47-bp long), which shows strong homology with the 5' end of several integrons from Gram-negative bacteria (Hall *et al.,* 1994; Rådström *et al.,* 1994) and Tn*5053* (Kholodii *et al.,* 1995), was discovered in the vicinity of the originally identified copy of IS*1415* in *R. erythropolis* NI86/21 (Nagy *et al.,* 1997b). However, the essential components of a typical integron, namely an *int* gene (encoding a site-specific recombinase of the λ integrase family, called integrase) and the linked *attI* site (recognized by Int) (Hall, 1997) for integration of incoming cassettes, are not evident. Analysis of the 22 kb DNA region downstream of this motif, which contained the cobalamin biosynthesis genes (De Mot *et al.,* 1994a); herbicide degradation genes (Nagy *et al.,* 1995a, b) and 20S proteasome structural genes (Tamura *et al.,* 1995; Nagy *et al.,* 1997c), did not detect any transposon-like structures. It is notable that genetic determinants (second proteasome operon and the thiocarbamate degradation genes) which are located downstream of the original 47-bp element of *R. erythropolis* NI86/21, are not found in other members of *Rhodococcus* species tested to date (Nagy *et al.,* 1995b; Lupas *et al.,* 1997). However, two more identical 47-bp elements have been identified, which were not associated with IS*1415*. These sequences constituted the 5' end of two identical DNA segments of unknown function. In close proximity to each of them was situated a site-specific recombinase gene of the Tn*3* resolvase family (Stark *et al.,* 1992). Although these recombinases differed from each other. The analysis of the function of these structures (recombinase and associated 47 bp sequence) may provide valuable information about organization and plasticity of *Rhodococcus* genome.

Gene Organization and Regulation

A significant number of *Rhodococcus* genes have been cloned and analyzed to date. In particular, those encoding pathways for the degradation of aromatic and other xenobiotic compounds have been studied extensively. Sequences of several operon-like structures encoding degradation of biphenyl (Masai *et al.*, 1995), isopropylbenzene (Kesseler *et al.*, 1996), dibenzothiophene desulphurization (Denome *et al.*, 1994), 3-hydroxyphenyl propionic acid (Barnes *et al.*, 1997), plant virulence loci (Crespi *et al.*, 1994) and other are now available. The overlapping of the coding regions in such structures appears to be quite common (see Kesseler *et al.*, 1996; Denome *et al.*, 1994; De Mot *et al.*, 1994a,b). The most frequently occurring overlap is presented by sequences ATGA and GTGA (where ATG and GTG are start codons used). A much longer overlap (of 17 bp) was found between two genes *hox*H and *hox*W belonging to the gene cluster of the soluble hydrogenase of *R. opacus* MR11 (Grzeszik *et al.*, 1997). These overlaps suggest transcriptional coupling of the genes involved. The dependence of enzyme expression in *Rhodococcus* on specific inducers (Desomer *et al.*, 1992; Komeda *et al.*, 1996b; Pogorelova *et al.*, 1996) or repressors (Li *et al.*, 1996) was demonstrated. A number of regulatory genes were found in rhodococci. The corresponding proteins have been assigned to various families of regulators (Table 1); however, it is notable that there are no known *Rhodococcus* members of such widely distributed families as LysR. Further analysis of rhodococci genes may however expand our knowledge about diversity of the associated regulatory elements.

Putative promoter regions were identified in a number of *Rhodococcus* operon-like structures. Interestingly, putative -10 and -35 promoter regions similar to those of *E. coli* were found to be rather common in *Rhodococcus* (Candidus *et al.*, 1994; Barnes *et al.*, 1997; Greszik *et al.*, 1997). This finding is in agreement with the analysis of *Streptomyces*, where a group of *E. coli* -like promoter sequences was also described (Hopwood *et al.*, 1986; Strohl, 1992). In a number of *Rhodococcus* promoter (Desomer *et al.*, 1992; Komeda *et al.*, 1996b) and putative regulatory regions (Kulakova *et al.*, 1997) sequences including inverted as well as direct repeats were identified. These were supposed to be essential for regulation of various *Rhodococcus* genes and in a case of the nitrilase gene an inverted repeat in promoter region were suggested to be a binding site for NitR regulator protein (Komeda *et al.*, 1996c). Easily identifiable ribosome binding sites (Shine-Dalgarno sequences) were found to precede most of the analyzed *Rhodococcus* genes (Komeda *et al.*, 1996b; Andersen *et al.*, 1997). In many cases those were similar to conserved *Streptomyces* sequence GGAGG (Strohl, 1992).

Table 1. Types of regulator proteins found in *Rhodococcus*.

Family	Regulator	Regulated function, catabolism	Strain	Reference
AraC/XylS	ThcR	thiocarbanates/ atrazine	*R. erythropolis* NI86/21	Nagy *et al.* 1995
AraC/XylS	FasR	fascination, virulence	*R. fascians* D188	Vereecke 1997
AraC/XylS	NitR	nitriles (nitrilase)	*R. rhodochrous* J1	Komeda *et al.* 1996a
AmiC	NhhC	nitriles (HM[a] hydratase)	*R. rhodochrous* J1	Komeda *et al.* 1996b
AmiC	NhlC	nitriles (LM[b] hydratase)	*R. rhodochrous* J1	Komeda *et al.* 1996c
Ara/Xyl ?	NhhD	nitriles (HM hydratase)	*R. rhodochrous* J1	Komeda *et al.* 1996b
MerR	NhlD	nitriles (LM[b] hydratase)	*R. rhodochrous* J1	Komeda *et al.* 1996c
IclR	HppR	3(3-hydroxyphenyl) propionate	*R. globerulus* PWD1	Barnes *et al.* 1997
IclR ?	CatR	catechol	*R. erythropolis* 1CP	Eulberg *et al.* 1997

[a] HM, high-molecular-mass; [b] LM, low-molecular-mass

An interesting feature of the *Rhodococcus* genome has emerged recently. Although, typical for most bacteria, it seems that the clustering of the genes belonging to the same biochemical pathways is not always applicable. Mapping and sequencing of *Rhodococcus* sp NCIMB12038 genes encoding naphthalene degradation revealed that only determinants encoding catalytic components of naphthalene dioxygenase (α and ß subunits, encoded by *nar*Aa and *nar*Ab genes) and *cis*-naphthalene dihydrodiol dehydrogenase (*narB*) were located in the same region of the bacterial genome (Larkin *et al.*, 1999; Kulakov *et al.*, 2000). Genes coding for the electron transfer components of the naphthalene dioxygenase and further enzymes of the naphthalene degradation pathway were not found in the same region (Kulakov *et al.*, unpublished).

Despite a significant number of analyzed *Rhodococcus* genes, relatively little research dealt directly with gene regulation in this group of microorganisms. One study has dealt with the transcriptional regulation of the *R. rhodochrous* J1 nitrilase (*nit*A) gene (Komeda *et al.*, 1996b). Authors of this study demonstrated that transcription of the *nit*A gene was inducible by isovaleronitrile, and that for the isovaleronitrile dependent induction regulatory gene encoding positive transcriptional regulator NitR was essential. This regulator gene was shown to be a member, although only distantly related, of XylS/AraC family of positive regulator proteins, which are involved in the regulation of various bacterial genes (Callegos *et al.*, 1993). The transcription start site (tsp) of the nitrilase gene was identified 26 bp upstream from ATG initiation codon and further 35 bp upstream -35 and -10 regions were found with similarity to *Streptomyces* consensus promoter sequence (Strohl, 1992).

The dibenzothiophene desulphurization (*dsz*) gene cluster, which was also analyzed in detail, did not show inducibility, but was strongly repressed by sulfate and sulfur-containing amino acids (Li *et al.*, 1997). Three regions, which influenced expression and repression of these genes were identified upstream of *dsz* gene cluster. Putative -10 and -35 promoter regions were also identified were shown to be similar to *Bacillus subtilis* promoters that utilize σ^B factor (Li *et al.,* 1996).

It is obvious that the study of *Rhodococcus* gene regulation is now at its very beginning. Regulatory regions (most of which are putative) identified so far may be considered as essentially similar to those of other eubacteria and especially *Streptomycetes*. However, only in few cases has functionality of the promoter sequences been confirmed by mutational analysis (Li *et al.*,

1996) or truncation (Komeda *et al.*, 1996b; Desomer *et al.*, 1992). In this respect it is worth noting that there are no published works dealing with the analysis of *Rhodococcus* RNA polymerase core enzymes and sigma factors. Considerable work will be required for understanding of gene regulation in rhodococci which will involve analysis of positive regulators (of which only NitR has been identified so far), repressors and their binding sites. Regulatory proteins found in *Rhodococcus* species are presented in Table 1.

Bacteriophages

There are few studies on the bacteriophages specific to the members of the genus *Rhodococcus*. However, the existence of *Rhodococcus* phages is long established (Prayser, 1976) and they seem do not have striking differences from the other bacterial viruses. Lysogeny was described for *R. erythropolis* infected with actinophages ϕC and ϕEC. These phages existed in lysogenic strains in an integrated as well as nonintegrated (plasmid-like) form and showed the ability to be transferred in bacterial crosses (Brownell and Denniston, 1984). Bacteriophage EC has been partially characterized and its use for the cloning has been discussed (Brownell *et al.*, 1980, 1982). Actinophages A and B were also found in *R. australis* CSIR-A201 and *R. equi* CSIR-A655 (Hiddema *et al.*, 1985). Bacteriophage Q4 capable of efficient general transduction in *R. erythropolis* was described (Dabbs, 1987; Dabbs and Sole, 1988) and the temperate phage NJL isolated from *R. rhodochrous* has been characterized by Sunairi *et al.* (1993).

Genetic Exchange and Recombination

Gene transfer and subsequent recombination in *Rhodococcus* was demonstrated in the early 1960's between *Nocardia erythropolis* and *Nocardia canicruria* (now both classified as *Rhodococcus erythropolis*) (Adams and Bradley, 1963; Adams, 1964). From the beginning, differences from the other well-studied systems (i.e. *E. coli*) became apparent. When recombinations in different *Rhodococcus* strains were analyzed, crosses between mutants of the same strains were infertile (self-incompatible) and only mating of different strains produced recombinants. However, recombinants arose, that were capable of backcrossing with both parental types which suggested that mating factors might be encoded by several different loci (Adams, 1964). Further experiments uncovered that mating types in the above mentioned strains were indeed determined by the multiple genes *matC*, *matE* (Brownell and Adams, 1968; Brownell and Kelly, 1969) and *matM* (Brownell, 1974). Further information concerning genes encoding self-incompatibility and conjugation came from the analysis of the 60-80 kbp plasmid pDA20 of *R. erythropolis* ATCC12674 and its deleted derivative pDA21 which encode resistance to arsenate, arsenite and chloramphenicol. Genes essential for chromosomal conjugation were located on an unstable genetic element and in this element two further regions were identified. One, 8 kbp, region encoded self-incompatibility phenotype and the other 3 kbp region was essential for the conjugation (Gowan and Dabbs, 1994). The exact function of all these loci has yet to be determined, and this will probably involve sequencing analysis of the above mentioned regions of the *R. erythropolis* genome and analysis of the mechanisms by which transfer of the *Rhodococcus* genetic material occur. Conjugational transfer of the linear plasmid pRHL2 was shown to take place between its host strain RHA1 and its derivatives. This transfer occurred with a frequency of 7.5×10^{-5} transconjugant per recipient cell, which suggests that linear plasmids possibly play a significant part in the horizontal transfer of biodegradative genes in rhodococci (Shimizu *et al.*, 2001).

Despite our incomplete understanding of the mechanisms involved in the gene transfer of various *Rhodococcus* genes it now seems certain that bacteria belonging to this genus possesses an enormous capacity to disseminate its genes. The haloalkane dehalogenase gene (*dhaA*) initially discovered in *Rhodococcus rhodochrous* NCIMB13064 (Kulakova *et al.*, 1997) exemplifies this. The *dhaA* gene was found in a number of *Rhodococcus* strains and even in *Pseudomonas* species isolated from contaminated sites in Europe, Japan and the United States (Poelarends *et al.*, 2000a). In all of the analyzed haloalkane degrading strains almost identical sequences of *dhaA* were found (Poelarends *et al.*, 2000 a, b). Although, in prototype strain NCIMB13064 *dhaA* was found on a 100 kb circular plasmid (Kulakova *et al.*, 1997) further analysis proved that various chloroalkane degraders can carry this gene both on the plasmids and chromosome (Poelarends *et al.*, 2000 b). An interesting variant of *dhaA* gene was found in *Mycobacterium* sp GP1. Dehalogenase ($DhaA_f$) encoded by this gene was almost identical to DhaA apart from additional 14 amino acid residues at the C-terminus. These additional residues were apparently originated from the fusion of *dhaA* with the distal part of *hheB* (haloalcohol dehalogenase gene of *Corynebacterium* sp N-1074) (Poelarends *et al.*, 1999).

Another feature of the recombination in *Rhodococcus* noted earlier was the absence of high-level DNA/DNA homology between recombining strains

(Kasweck and Little, 1982). Although the overall DNA homology between compatible strains of *Rhodococcus* and *Nocardia* can be low, as determined by DNA/DNA reassociation (Clark and Brownell, 1972; Kasweck and Little, 1982), regions of extensive homology do exist. When DNA from *R. erythropolis* mating type cE2 was hybridized with DNA from the compatible mating type Ce3 the level of homology detected was only approximately 60%. However, when the reverse DNA/DNA reassociation was done, nearly 100% homology was shown. These experiments suggested that the Ce3 strain had a considerable amount of DNA with a limited level of homology to that of cE2. Taking into account the number of aberrant segregants observed in recombination studies (Adams and Brownell, 1976), the presence of large linear and circular plasmids in these bacteria and the possibility of a linear chromosome the low level of DNA/DNA homology between different strains may reflect flexibility of the *Rhodococcus* genome.

Instability of the *Rhodococcus* Genome

The results described above imply the presence of significant differences in the genomes of closely related *Rhodococcus* strains. However, the nature of these differences remains to be discovered. It is possible that large regions of bacterial chromosome can be deleted, or that differences in the DNA/DNA homology between strains may be explained by their variable plasmid content.

Genetic instability is well documented for *Streptomyces* species (Leblond *et al.*, 1990; Volff *et al.*, 1993). Phenotypic traits connected with this phenomenon vary in different species and can include morphological and physiological differentiation, which is very complex in streptomycetes. However, there are several general features associated with genetic instability in *Streptomyces*. Mutants that arise usually exhibit genomic rearrangements such as deletions and amplifications of certain DNA sequences. Some deletions were shown to be extremely large (Flett and Cullum, 1987; Birch *et al.*, 1989) and can affect about a quarter of the genome without reducing the survival of the bacteria (Leblond *et al.*, 1991). Another typical feature of the genetic instability in *Streptomyces* is the presence of amplifiable units of DNA (AUD) which can be tandemly reiterated up to several hundred times (Fishman and Hersberger, 1983; Hornemann *et al.*, 1989). Functional ORFs were found to be located in the unstable region on AUD fragment (Schneider *et al.*, 1993) and involvement of DNA gyrase and a SOS -like system in genetic instability was suggested (Volff *et al.*, 1993). No genomic deletions and rearrangements of this type have been described so far for *Rhodococcus*, however few studies were conducted in this area.

The potential flexibility of the *Rhodococcus* genome becomes clear when homologous recombination between cloned sequences and those present in the bacterial genome is used to assign functions to particular genes. Barnes *et al.* (1997) attempted to "knock down" chromosomal 3-hydroxyphenylacetate genes (*hpp*) of *Rhodococcus globerulus* PWD1 with a construct in which the kanamycin resistance marker from pUC4K (Vieira and Messing, 1982) was inserted in the cloned *hpp* genes. All of the 1600 Km^R mutants screened appeared to be the result of non-homologous recombination and 150 were analyzed by PCR and hybridization to confirm this. The plasmids integrated at multiple sites at various locations in the chromosome, despite having more then 500 bp of sequences homologous to the target on the either side of the kanamycin marker in the constructs. Similarly multiple insertions of the same kanamycin resistance marker into non-homologous regions have been observed when attempts were made to interrupt extradiol dioxygenase genes putatively involved in naphthalene and isopropyl benzene catabolism. However, in these experiments about 10% of recombinants resulted from the homologous recombination with the two of the three targeted genes (Irvine *et al.*, 2000).

In contrast, some gene interruption experiments with *Rhodococcus* are much more sequence specific. Masai *et al.* (1995) also employed constructs in which the kanamycin resistance gene from pUC4K was inserted into the biphenyl catabolic genes (*bph*) cloned from plasmids pRHA1 and pHRA2 of *Rhodococcus* strain RHA1. These constructs when introduced into original strain successfully recombined with the homologous wild type genes and the inactivation of the functions proved they belonged to the biphenyl catabolism operon. Similar specificity was achieved during inactivation of the biphenyl/chlorobiphenyl genes (*bpd*) in *Rhodococcus* strain M5 using a kanamycin resistance marker (Wang *et al.*, 1995).

A more detailed study of illegitimate integration of plasmid DNA in *Rhodococcus fascians* was undertaken by Desomer et al. (1991). After transformation of bacterial cells with plasmids pRF41 (derivative of pUC13) or related plasmid pRF32, illegitimate integrations always occurred at a palindrome sequence CCGCGG present in one copy in pRF41 and two copies in pRF32. In this case a 5-base pair sequence (CCGCA) was also found almost immediately upstream of the palindrome, which could played a role in the integration. These workers postulated that a linear recombination intermediate was generated by cleavage of the plasmid at the specific site. This was thought to involve an enzyme similar to a restriction endonuclease but no such activity was detected. Surprisingly, plasmid insertion in the genome appeared to occur at random, non-homologous sequences leading to a wide range of mutant phenotypes. In some cases a small target site deletion (12 bp) occurred (Desomer *et al.*, 1991).

Genome rearrangements leading to decryption of a silent biodegradation pathway have been discovered in *Rhodococcus rhodochrous* NCIMB13064. This strain was isolated as utilizing a wide range of 1-haloalkanes as sole carbon and energy source (Curragh *et al.*, 1994) and it was unable to grow on aromatic compounds such as naphthalene. Plasmids pRTL1(100 kb) and pRTL2 (80 kb) were found in NCIMB13064 (Kulakova *et al.,* 1995). The haloalkane dehalogenase gene (*dhaA*), encoding dehalogenation of short chain 1-haloalkanes (chain length C_3 - C_{10}) is located on the pRTL1 plasmid (Kulakova *et al.*, 1997). Analysis of the NCIMB13064 strain involved isolation of Dha^- derivatives. Amongst these derivatives two strains, P200 and P400 were found which acquired the ability to utilize naphthalene (Nah^+) as sole carbon and energy source (Kulakova *et al.*, 1996). Both strains had lost the ability to degrade short chain 1-haloalkanes (Dha^-) but retained the ability to use long chain (C_{12} - C_{18}) 1-haloalkanes. In derivative strain P200 a 20 kbp fragment of pRTL1 was shown to be integrated into bacterial chromosome (Kulakova *et al.*, 1995). Strain P400 had lost both pRTL1 and pRTL2 plasmids which were detected in the original NCIMB13064 strain. The Nah^+ phenotype in these derivatives appeared to be fairly stable, however, Dha^+ revertants of the strain P400, which arose at frequency of about 10^{-8}, had lost the ability to utilize naphthalene, were Dha^+ and had regained both original plasmids (Kulakova *et al.*, 1996). It is interesting to note that insertion sequence IS*2112* (see above) was found 5 kbp upstream of the *dha*A gene. Although this element does not appear to comprise a composite transposon with the *dha*A gene it seems to be involved in genome rearrangements in NCIMB13064. IS*2112* appeared to change its location as a result of transposition and as a result of other rearrangements of the NCIMB13064 genome (Kulakov *et al.*, 1999).

Research to date suggests that there is considerable instability in *Rhodococcus* genome that may be directly associated with the remarkable biodiversity of this group of bacteria. The possible role of mobile genetic elements coupled with major genomic rearrangements associated with global regulation and gene expression in *Rhodococcus* remains to be investigated.

Evolution of the *Rhodococcus* Genome

Although, information regarding the organization of the *Rhodococcus* genome remains rather fragmented, nevertheless some conclusions concerning the evolution of particular genes or gene clusters may be reached using the data available. One of the most analyzed classes of *Rhodococcus* genes are the extradiol dioxygenase genes, which encode enzymes participating in the degradation of a wide range of aromatic pollutants. A number of *Rhodococcus* extradiol dioxygenase genes have been characterized recently. Their functional and phylogenetic studies allow some conclusions to be drawn about the biodegradative diversity of this group of microorganisms.

Ring cleavage is an important step in an aerobic degradation of aromatic compounds by bacteria. The enzymes belonging to one of the two different classes catalyze this: intradiol or extradiol dioxygenases (Harayama *et al.*, 1992). Intradiol dioxygenase genes participating in catechol degradation by *Rhodococcus* were analyzed recently (Eulberg *et al.*, 1997). Extradiol dioxygenases which cleave aromatic ring between one hydroxylated and another adjacent nonhydroxylated carbon are key enzymes in many bacterial pathways for the degradation of aromatic compounds (Harayama *et al.*, 1992). A large number of genes encoding extradiol dioxygenases were studied and the crystal structures of two *Pseudomonas* 2,3-dihydroxybiphenyl dioxygenases were resolved (Han *et al.*, 1995; Senda *et al.*, 1996). A comprehensive classification of extradiol dioxygenases based on structure validated phylogenetic analysis was proposed by Eltis and Bolin (1996). According to that classification all extradiol dioxygenases could be divided into two major unrelated types with the majority of the known enzymes belonging to type I. The type I enzymes display different degrees of relatedness and they are divided into five families and a number of subfamilies. The majority of the characterized extradiol dioxygenases belong to *Pseudomonas* spp. and other Gram-negative microorganisms. However, a significant number of *Rhodococcus* encoded extradiol dioxygenases have been also characterized. Enzymes of both types were found in *Rhodococcus* species (Barnes *et al.*, 1997; Kulakov *et al.*, 1998) with the majority belonging to family 3 of the type I. This bias towards one family may be explained by the fact that most analyzed extradiol dioxygenases of *Rhodococcus* are from biphenyl degradation pathways. The phylogenetic relationships of family 3 dioxygenases are presented in Figure 1. It should be noted that *Rhodococcus* extradiol dioxygenases belong to the same subfamilies (3.B - 3.E) as *Pseudomonas* ones. These observation may indicate that extensive exchange

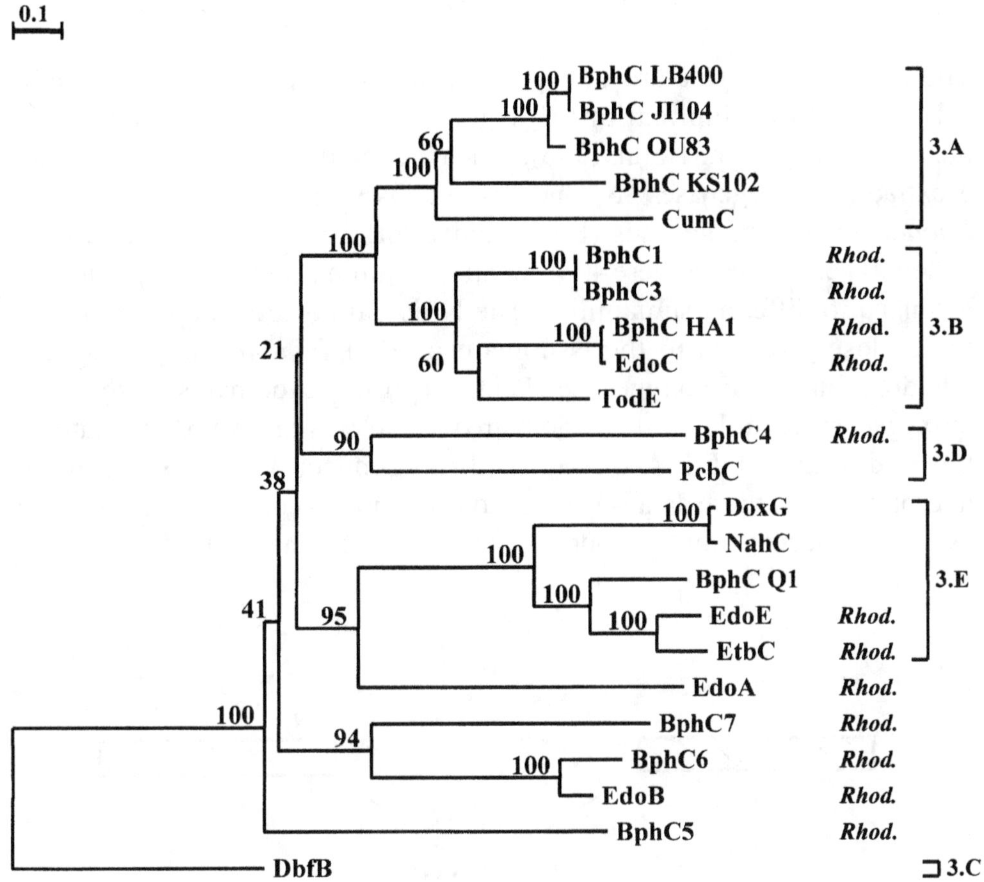

Figure 1. Phylogenetic tree for type I, family 3 extradiol dioxygenases obtained by the neighbor-joining approach. Bootstrap values (%) are indicated at the nodes. Sub-families, 3.A - 3.D given according to Eltis and Bolin (1996). BphC_LB 400, *P. cepacia* LB400, 2,3-dihydroxybiphenyl dioxygenase (accession No X66122); BphC JI104, *P. aeruginosa* JI104 dioxygenase (D83056); BphC OU83 *P. putida* OU83 2,3-dihydroxybiphenyl dioxygenase (X91876); BphC_KS102, *Pseudomonas sp.* KKS102 2,3-dihydroxybiphenyl dioxygenase (M26433); CumC, *P. fluorescens* IPO1 catechol 2,3-dioxygenase (D37828) ; TodE, *P. putida* F1 3-methylcatechol 2,3-dioxygenase (J04996) [16]; BphC HA1, *Rhodococcus sp.* RHA1 2,3-dihydroxybiphenyl dioxygenase (D32142); BphC1, *R. globerulus* P6, 2,3-dihydroxybiphenil dioxygenase I(3) (X75633); DbfB, *Sphingomonas paucimobilis* RW1, 2,2',3-trihydroxybiphenyl dioxygenase (X72850); EtbC, *Rhodococcus sp.* RHA1, meta- ring cleavage dioxygenase (M76438); BphC Q1, *P. paucimobilis* Q1, 2,3-dihydroxybiphenyl dioxygenase (M20640); NahC, *P. putida* PpG7, 1,2-dihydroxynaphthalene dioxygenase (J04994); PcbC, *Pseudomonas sp.* DJ12, 2,3-dihydroxybiphenyl dioxygenase (D44550); DoxG, *Pseudomonas sp.* C18, 1,2-dihydroxynaphthalene dioxygenase (M60405); EdoA, EdoB, EdoC, *Rhodococcus rhodochrous* NCIMB13064 extradiol dioxygenases (Kulakov et al., 1998); BphC3 - BphC7 *Rhodococcus erythropolis* TA421 2,3-dihydroxybiphenyl 1,2-dioxygenases (D88015-D88019).

of dioxygenase genes between *Pseudomonas* and *Rhodococcus* took place in the course of evolution of family 3 extradiol dioxygenases.

The two domain structure of type I extradiol dioxygenases supposedly originated from duplication of a gene encoding one domain proto-dioxygenase (Eltis and Bolin, 1996). When the comparative analysis of family 3 extradiol dioxygenases' N- and C-domains was performed, two of the *Rhodococcus* dioxygenases (EdoA and EdoB) showed an extraordinary pattern. These enzymes were 40 % identical at amino acid level and evidently belonged to different subfamilies. The N-domain of EdoB appeared to be more closely related to the N-domain of *Pseudomonas sp.* DJ12, 2,3-dihydroxybiphenyl dioxygenase (PcbC) than the C-domains of the same dioxygenases. Similarily the C-domain of PcbC was more closely related to the C-domain of EdoA extradiol dioxygenase. The most probable interpretation of these data is that a recombination event resulting in domain exchanges occurred in the course of evolution leading to the formation of

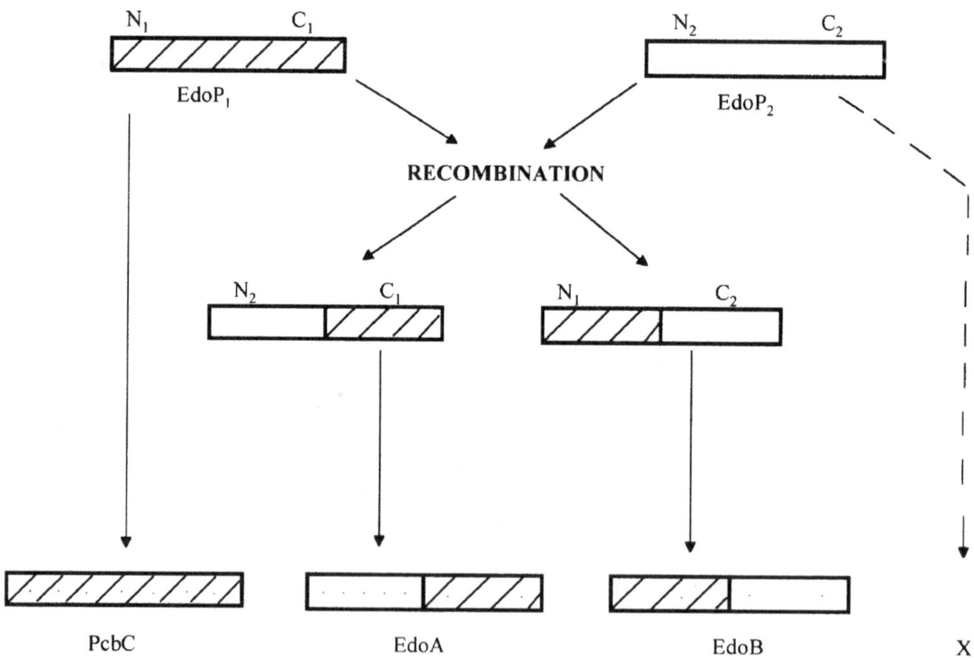

Figure 2. Hypothetical origin of two *Rhodococcus* (EdoA and EdoB) and one *Pseudomonas* (PcbC) extradiol dioxygenases. $EdoP_1$ and $EdoP_2$ are hypothetical ancestral extradiol dioxygenases. PcbC evolved directly from $EdoP_1$. Recombination between $edoP_1$ and $edoP_2$ genes gave hybrid species evolution of which resulted in EdoA and EdoB extradiol dioxygenases. The fate of $EdoP_2$ lineage is unknown.

the extant extradiol dioxygenases EdoA, EdoB and PcbC. However, the possibility of parallel evolution in the N- and C- domains of these dioxygenases cannot be completely excluded. Schematic presentation of a possible evolution of PcbC, EdoA and EdoB dioxygenases is presented in Figure 2. The possibility of "domain swapping" in the evolution of extradiol dioxygenases has been considered previously. For example *P. putida* MT-2 catechol 2,3-dioxygenase XylE was thought to have originated from domain fusion involving catechol 2,3-dioxygenase NahH ancestor (Harayama and Rekik, 1989; Benjamin *et al.*, 1991). Current analyses provides evidence of possible domain swapping involving enzymes belonging to *Pseudomonas* and *Rhodococcus* species, which implies natural exchange of genetic information between these phylogenetically different groups of microorganisms. Although the common origin of all type I extradiol dioxygenases was suggested earlier by Eltis and Bolin (1996) these data imply that exchange of genetic information between Gram-negative and Gram-positive species played an important part in the evolution of at least one group of these enzymes. The involvement of the genetic recombination in formation of *dhhA* gene region is discussed in the previous section.

Conclusions

Although, the genus *Rhodococcus* includes many environmentally and industrially important species the organization of the *Rhodococcus* genome is poorly understood. Sequencing information is limited to gene and gene clusters encoding various pathways and a few isolated plasmids. Great variability of rhodococci and remarkable instability of many of their characteristics makes genomic studies of the genus especially difficult. Several particular features may be pointed out as likely to be characteristic of the *Rhodococcus* genome: *i*) the presence of large linear plasmids and possibly a linear chromosome; the termini of linear plasmids show essential similarity to those of *Streptomyces*, although some important differences have been observed; *ii*) circular plasmids are also common in *Rhodococcus* species; *iii*) a great variety of biodegradative genes have been found in rhodococci, and the operon–like organization of some of these determinants may be in question; *iv*) horizontal transfer of the genetic material has apparently played an essential role in formation of the *Rhodococcus* genome; *v*) instability of various *Rhodococcus* characteristics is well known, however this is not supported by a detailed analysis of genomic instability as was done for some *Streptomyces* species.

Rhodococcus genomics is clearly at an early stage in its development. However, the general availability of a completed genome sequence will be only the first step in elucidating the genomic structure of these diverse bacteria. It has only recently become clear that there is a wide diversity of plasmids dispersed amongst strains of this genus. Many of these are very large linear plasmids that comprise a significant proportion of the genome and undoubtedly contribute to a wide strain variation in genome size. The full extent of the diversity and distribution of such plasmids is not known along with extent of inter-plasmid and genome-plasmid recombination. However, it is likely that there is considerable variation in gene organization between strains of this genus that only a comparative genome study can elucidate. In our opinion a detailed study of the genomes of the rhodococci will raise many fundamental questions regarding the organization of bacterial genomes.

References

Adams, J.N. 1964. Recombination between *Nocardia erythropolis* and *Nocardia canicruria*. J. Bacteriol. 88: 865-876

Adams, J.N., and Bradley, S.G. 1963. Recombination events in the bacterial genus *Nocardia*. Science. 140: 1392-1394.

Adams, J.N., and Brownell, G.H. 1976. Genetic studies in *Nocardia erythropolis*. In: The Biology of the *Nocardiae*. Goodfellow, M., Brownell, G.H., and Serrano, J.A., eds. Academic Press, London. p 285-309.

Andersen, S.J., Quan, S., Gowan, B., and Dabbs, E.R. 1997. Monooxygenase-like sequence of a *Rhodococcus equi* gene conferring increased resistance to rifampin by inactivating this antibiotic. Antimicrob. Agents. Chem. 41: 218-221.

Bardarov, S., Kriakov, J., Carrierre, C., Yu S., Vaamonde, C., McAdam, R.A., Bloom, B.R., Hatfull, G.F., and Jacobs, Jr W.R. 1997. Conditionally replicating mycobacteriophages: a system for transposon delivery to *Mycobacterium tuberculosis*. Proc. Natl. Acad. Sci. USA. 94: 10961-10966.

Barnes, M.R., Duetz, W.A., and Williams, P.A. 1997. A 3-(3-hydroxyphenyl)propionic acid catabolic pathway in *Rhodococcus globerulus* PWD1: cloning and characterization of the *hpp* operon. J. Bacteriol. 179: 6145-6153.

Benjamin, R.C, Voss, J.A., and Kunz, D.A. 1991. Nucleotide sequence of *xylE* from TOL pDK1 plasmid and structural comparison with the isofunctional catechol 2,3-dioxygenase genes from TOL pWW0 and NAH7. J. Bacteriol. 173: 2724-2728.

Birch, A., Häusler, A., Vögtli, M., Krek, W., and Hütter, R. 1989. Extremely large chromosomal deletions are intimately involved in genetic instability and genomic earrangements in *Streptomyces glaucescens*. Mol. Gen. Genet. 217: 447-458.

Brownell, G. H. 1974. A new nocardial mating strain. Am. Soc. Microbiol. abstract p. 41.

Brownell, G.H., and Adams, J.N. 1968. Linkage and segregation of a mating type specific phage and resistance chararacters in nocardial recombinants. Genetics. 60: 437-48

Brownell, G.H., and Denniston, K. 1984. Genetics of the nocardioform bacteria. In: The Biology of the Actinomycetes. Goodfellow, M., Mordarski, M., and Williams, S.T. eds. Academic Press, New York. P. 201-208.

Brownell, G.H., Enquist, L.W., and Denniston-Thompson, K. 1980. An analysis of the genome of the actinophage ϕEC. Gene. 12: 311-314.

Brownell, G.H., and Kelly, K.L. 1969. Inheritance of mating type factors in nocardial recombinants. J. Bacteriol. 99: 25-36.

Brownell, G.H., Saba, J.A., Denniston-Thompson, K., and Enquist, L.W. 1982. The development of a *Rhodococcus*-actinophage cloning system. Dev. Ind. Microbiol. 23: 287-298.

Brunhuber, N.M., Banerjee, A., Jacobs, Jr. W.R., and Blanchard, J.S. 1994. Cloning, sequencing, and expression of *Rhodococcus* L-phenylalanine dehydrogenase. Sequence comparisons to amino-acid dehydrogenases. J. Biol. Chem. 269: 16203-16211.

Candidus, S., Vanpee, K.H., and Lingens, F. 1994. The catechol 2,3-dioxygenase gene of *Rhodococcus rhodochrous* CTM - nucleotide-sequence, comparison with isofunctional dioxygenases and evidence for an active-site histidine. Microbiology. 140: 321-330.

Clark, J.E., and Brownell, G.H. 1972. Genophore homologies among compatible nocardiae. J. Bacteriol. 109: 720-729.

Crespi, M., Messens, E., Caplan, A.B., Vanmontagu, M., and Desomer, J. 1992. Fasciation induction by the phytopathogen *Rhodococcus fascians* depends upon a linear plasmid encoding a cytokinin synthase gene. EMBO Journal. 11: 795-804.

Crespi, M., Vereecke, D., Temmerman, W., Vanmontagu, M., and Desomer, J. 1994. The *fas* operon of *Rhodococcus fascians* encodes new genes required for efficient fasciation of host plants. J. Bacteriol. 176: 2492-2501.

Curragh, H., Flynn, O., Larkin, M.J., Stafford, T.M., Hamilton, J.T.G., and Harper, D.B. 1994. Haloalkane degradation and assimilation by *Rhodococcus rhodochrous* NCIMB13064. Microbiology. 140: 1433-1442.

Dabbs, E.R. 1987. A generalized transducing bacteriophage for *Rhodococcus erythropolis*. Mol. Gen. Genet. 206: 116-120.

Dabbs, E.R., and Sole, G.J. 1988. Plasmid-borne resistance to arsenate, arsenite, cadmium and chloramphenicol in a *Rhodococcus* species. Mol. Gen. Genet. 211: 148-154.

De Mot, R., Nagy, I., Schoofs, G., and Vanderleyden, J. 1994a. Sequences of the cobalamin biosynthetic genes *cobK, cobL,* and *cobM* from *Rhodococcus* sp. NI86/21. Gene. 143: 91-93.

De Mot, R., Nagy, I., Schoofs, G., and Vanderleyden, J. 1994b. Sequence of a *Rhodococcus* gene cluster encoding the subunits of ethanolamine ammonia-lyase and an APC-like permease. Can. J. Microbiol. 40: 403-407.

De Mot, R., Nagy, I., De Schrijver, A., Pattanapipitpaisal P., Schoofs, G., and Vanderleyden, J. 1997. Structural analysis of the 6 kb cryptic plasmid pFAJ2600 from *Rhodococcus erythropolis* NI86/21 and construction of *Escherichia coli-Rhodococcus* shuttle vectors. Microbiology. 143: 3137-3147.

Denis-Larose, C., Bergeron, H., Labbe, D., Greer, C.W., Hawari, J., Grossman, M.J., Eltis, L.D., and Bolin, J.T. 1996. Evolutionary relationship among extradiol dioxygenases. J. Bacteriol. 178: 5930-5937.

Denis-Larose, C., Bergeron, H., Labbe, D., Greer, C.W., Hawari, J., Grossman, M.J., Sankey, B.M., and Lau, P.C. 1998. Characterization of the basic replicon of *Rhodococcus* plasmid pSOX and development of a *Rhodococcus-Escherichia* coli shuttle vector. Appl. Environ. Microbiol. 64: 4363-7.

Denis-Larose, C., Labbe, D., Bergeron, H., Jones, A. M., Gree,r C.W., Alhawari, J., Grossman, M.J., Sankey, B.M., and Lau, P.C.K. 1997. Conservation of plasmid-encoded dibenzothiophene desulfurization genes in several rhodococci. Appl. Environ. Microbiol. 63: 2915-2919.

Denome, S.A., Oldfield, C., Nash, L. J., and Young, K.D. 1994. Characterization of the desulfurization genes from *Rhodococcus* sp. strain IGTS8. J. Bacteriol. 176: 6707-6716.

Denome, S.A., and Young, K.D. 1995. Identification and activity of 2 insertion-sequence elements in *Rhodococcus* sp strain IGTS8. Gene. 161: 33-38.

Desomer, J., Crespi, M., and Vanmontagu, M. 1991. Illegitimate integration of non-replicative vectors in the genome of *Rhodococcus fascians* upon electrotransformation as an insertional mutagenesis system. Mol. Microbiol. 5: 2115-2124.

Desomer, J., Vereecke, D., Crespi, M., and Van Montagu, M. 1992. The plasmid-encoded chloramphenicol-resistance protein of *Rhodococcus fascians* is homologous to the transmembrane efflux proteins. Mol. Microbiol. 6: 2377-2385.

Eltis, L.D., and Bolin, J.T. 1996. Evolutionary relationship among extradiol dioxygenases. J. Bacteriol. 178: 5930-5937.

Eulberg, D., Golovleva, L.A., and Schlomann, M. 1997. Characterisation of catechol catabolic genes from *Rhodococcus erythropolis* 1CP. J. Bacteriol. 179: 370-381.

Ferretti, J.J., McShan, W.M., Adjic, D., Savic, D., Savic, G., Lyon, K., Primeaux, C., Sezate, S.S., Surorov, A.N., Kenton, S., Lai, H., Lin, S., Qian, Y., Jia, H.G., Najar, F.Z., Ren, Q., Zhu, H., Song, L., White, J., Yuan, X., Clifton, S.W., Roe, B.A., and McLaughlin, R.E. 2001. Complete genome sequence of an M1 strain of *Streptococcus pyogenes*. Proc. Natl. Acad. Sci. U.S.A. 98: 4658-4663.

Finnerty, W.R. 1992. The biology and genetics of the genus *Rhodococcus*. Ann. Rev. Microbiol. 46: 193-218.

Flett, F. and Cullum, J. 1987. DNA deletions in spontaneous chloramphenicol-sensitive mutants of *Streptomyces coelicolor* A3(2) and *Streptomyces lividans* 66. Mol. Gen. Genet. 207: 499-502.

Fishman, S.E., and Hershberger, C.L. 1983. Amplified DNA in *Streptomyces fradiae*. J. Bacteriol. 155: 459-466.

Goodfellow, M. 1989. Suprageneric classification of actinomycetes. In: Manual of Systematic Bacteriology Bergey's, Volume 4. Williams, S.T., Sharpe, M.E., and Holt, J.G. eds. Williams and Wilkins, Baltimore. p 2333-2339.

Goodfellow, M., Alderson, G., and Chun, J. 1998. Rhodococcal systematics: problems and developments. Antonie van Leeuwenhoek. 74: 3–20.

Gowan, B., and Dabbs, E.R. 1994. Identification of DNA involved in *Rhodococcus* chromosomal conjugation and self-incompatibility. FEMS Microbiol. Lett. 115: 45-50.

Green, E.P., Tizard, M.L., Moss, M.T., Thompson, J., Winterbourne, D.J., McFadden, J.J., and Hermon-Taylor, J. 1989. Sequence and characteristics of IS*900*, an insertion element identified in human Crohn's disease isolate of *Mycobacterium paratuberculosis*. Nucleic Acids Res. 17: 9063-9073.

Grzeszik, C., Lübbers, M., Reh, M., and Schlegel, H.G. 1997. Genes encoding the NAD-reducing hydrogenase of *Rhodococcus opacus* MR11. Microbiology. 143: 1271-1286

Guilhot, C., Otal, I., Van Rompaey, I., Martin, C., and Gicquel, B. 1994. Efficient transposition in mycobacteria: construction of *Mycobacterium smegmatis* insertional mutant libraries. J. Bacteriol. 176: 535-539.

Hall, R.M., Brown, H.J., Brookes, D.E., and Stokes, H.W. 1994. Integrons found in different locations have identical 5'-ends but variable 3'-ends. J. Bacteriol. 176: 6286-6294.

Han, S., Eltis, L.D., Timmis, K.N., Muchmore, S.W., and Bolin, J.T. 1995.

Crystal structure of the biphenyl-cleaving extradiol dioxygenase from PCB-degrading pseudomonad. Science. 270: 976-980.

Harayama, S., Kok, M., and Neidle, E.L. 1992. Functional and evolutionary relationships among diverse dioxygenases. Annu. Rev. Microbiol. 46: 565-601.

Harayama, S., and Rekik, M. 1989. Bacterial aromatic ring-cleavage enzymes are classified into two different gene families. J. Biol. Chem. 264: 15328-15333.

Hayakawa, T., Tanaka, T., Sakaguchi, K., Otake, N., and Yonehara, H. 1979. A linear plasmid-like DNA in *Streptomyces* sp. producing lankacidin group antibiotics. J. Gen. Appl. Microbiol. 25: 255-260.

Hiddema, R., Curran, M.D., Ferreira, N.P., Coetzee, J.N., and Lecatsas, G. 1985. Characterization of phages derived from strains of *Rhodococcus australis* and *R. equi*. Intervirology. 23: 109-111.

Hinnebusch, J., and Tilly, K. 1993. Linear plasmids and chromosomes in bacteria. Mol. Microbiol. 10: 917-922.

Hiraga, S-I., Sugiyama, T., and Itoh, T. 1994. Comparative analysis of the replicon region of eleven ColE2-related plasmids. J. Bacteriol. 176: 7233-7243.

Hopwood, D.A., Bibb, M.J., Chater, K.F., Janssen, G.R., Malpartida, F., and Smith, P. 1986. In: Regulation of Gene Expression. 25 Years on. Booth, J.R., and Higgins, C.F. eds. Symp. Soc. Gen. Microbiol., University of Cambridge Press, Cambridge, UK. pp. 251-276.

Hornemann, U., Otto, C.J., and Zhang, X.Y. 1989. DNA amplification in *Streptomyces achromogenes* subsp. *rubradiris* is accompanied by deletion, and the amplified sequences are conditionally stable and can be eliminated by two pathways. J. Bacteriol. 171: 5817-5822.

Irvine, V.A., Kulakov, L.A., and Larkin, M.J. 2000. The diversity of extradiol dioxygenase (*edo*) genes in cresol degrading rhodococci from creosote-contaminated site that express a wide range of degradative abilities. Antonie van Leeuwenhoek. 78: 341– 352.

Jäger, W., Schäfer, A., Kalinowski, J., and Pühler, A. 1995. Isolation of insertion elements from Gram-positive *Brevibacterium, Corynebacterium* and *Rhodococcus* strains using the *Bacillus subtilis sacB* gene as a positive selection marker. FEMS Microbiol. Lett. 126: 1-6.

Jäger, W., Schäfer, A., Pühler, A., Labes, G., and Wohlleben, W. 1992. Expression of the *Bacillus subtilis sacB* gene leads to sucrose sensitivity in the gram-positive bacterium *Corynebacterium glutamicum* but not in *Streptomyces lividans*. J. Bacteriol. 174: 5462-5465.

Kalkus, J., Dorrie, C., Fischer, D., Reh, M., and Schlegel, H.G. 1993. The giant linear plasmid pHG207 from *Rhodococcus* sp. encoding hydrogen autotrophy - characterization of the plasmid and its termini. J. Gen. Microbiol. 139: 2055-2065.

Kalkus, J., Menne, R., Reh, M., and Schlegel, H.G. 1998. The terminal structures of linear plasmids from *Rhodococcus opacus*. Microbiology. 144: 1271-1279.

Kasweck, K.L., and Little, M.L. 1982. Genetic recombination in *Nocardia asteroides*. J. Bacteriol. 149: 403-406.

Kesseler, M., Dabbs, E. R., Averhoff, B., and Gottschalk, G. 1996. Studies on the isopropylbenzene 2,3-dioxygenase and the 3- isopropylcatechol 2,3-dioxygenase genes encoded by the linear plasmid of *Rhodococcus erythropolis* BD2. Microbiology. 142: 3241-3251.

Kholodii, G.Y., Mindlin, S.Z., Bass, I.A., Yurieva, O.V., Minakhina, S.V., and Nikiforov, V.G. 1995. Four genes, two ends, and a *res* region are involved in transposition of Tn*5053*: a paradigm for a novel family of transposons carrying either a *mer* operon or an integron. Mol. Microbiol. 17: 1189-1200.

Kobayashi, M., Komeda, H., Shimizu, S., Yamada, H., and Beppu, T. 1997. Characterization and distribution of IS*1164* that exists in the high molecular mass nitrile hydratase gene cluster of the industrial microbe *Rhodococcus rhodochrous* J1. Proc. Japan. Acad. 73B: 104-108.

Komeda, H., Kobayashi, M., and Shimizu, S. 1996a. A novel gene-cluster including the *Rhodococcus rhodochrous* J1 *nhlBA* genes encoding a low-molecular-mass nitrile hydratase (l- nhase) induced by its reaction-product. J. Biol. Chem. 271: 15796-15802.

Komeda, H., Kobayashi, M., and Shimizu, S. 1996b. Characterization of the gene cluster of high molecular mass nitrile hydratase (H-NHase) induced by its reaction product in *Rhodococcus rhodochrous* J1. Proc. Natl. Acad. Sci. USA. 93: 4267-4272.

Komeda, H., Hori, Y., Kobayashi, M., and Shimizu, S. 1996c. Transcriptional regulation of the *Rhodococcus rhodochrous* J1 *nitA* gene encoding a nitrilase. Proc. Natl. Acad. Sci. USA. 93: 10572-10577.

Kulakova, A.N., Larkin, M.J., and Kulakov, L.A. 1997. The plasmid-located haloalkane dehalogenase gene from *Rhodococcus rhodochrous* NCIMB13064. Microbiology. 143: 109-115.

Kulakova, A.N., Stafford, T.M., Larkin, M.J., and Kulakov, L.A. 1995. Plasmid pRTL1 controlling 1-chloroalkane degradation by *Rhodococcus rhodochrous* NCIMB13064. Plasmid. 33: 208-217.

Kulakov, L.A., Allen, C.C.R., Lipscomb, D.A., and Larkin, M.J. 2000. Cloning and characterization of a novel *cis*-naphthalene dihydrodiol dehydrogenase gene (*narB*) from *Rhodococcus* sp. NCIMB12038. FEMS Microbiol. Lett. 182: 327-331.

Kulakov, L.A., Larkin, M.J., and Kulakova, A.N. 1997. Cryptic plasmid pKA22 isolated from the naphthalene degrading derivative of *Rhodococcus rhodochrous* NCIMB13064. Plasmid. 38: 61-69.

Kulakov, L.A., Poelarends, G.J., Janssen, D.B., and Larkin, M.J. 1999. Characterisation of IS*2112*, a new insertion sequence from *Rhodococcus*, and its relationship with mobile elements belonging to the IS*110* family. Microbiology. 145: 561-568.

Lang, S., and Philp, J. 1998. Surface –active lipids in rhodococci. Antonie van Leeuwenhoek. 74: 59-70.

Larkin, M.J., Allen, C.C.R., Kulakov, L.A., and Lipscomb, D.A. 1999. Purification and characterization of a novel naphthalene dioxygenase from *Rhodococcus* sp strain NCIMB12038. J. Bacteriol. 181: 6200-6204.

Larkin, M.J., De Mot, R., Kulakov, L.A., and Nagy, I. 1998. Applied aspects of *Rhodococcus* genetics. Antonie van Leeuwenhoek. 74: 133-153.

Leblond, P., Demuyter, P., Simonet, J.-M., and Decaris, B. 1990. Genetic instability and hypervariability in *Streptomyces ambofaciens*: towards an understanding of a mechanism of genome plasticity. Mol. Microbiol. 4: 707-714.

Leblond, P., Demuyter, P., Simonet, J.-M., and Decaris, B. 1991. Genetic instability and associated genome plasticity in *Streptomyces ambofaciens*: pulsed field gel electrophoresis evidence for large DNA alterations in a limited genomic region. J. Bacteriol. 173: 4229-4233.

Leskiw, B.K., Mevarech, M., Barritt, L.S., Jensen, S.E., Henderson, D.J., Hopwood, D.A., Bruton, C.J., and Chater, K.F. 1990. Discovery of an insertion sequence, IS*116*, from *Streptomyces clavuligerus* and its relatedness to other transposable elements from actinomycetes. J. Gen. Microbiol. 136: 1251-1258.

Li, M.Z., Squires, C.H., Monticello, D.J., and Childs, J.D. 1996. Genetic analysis of the *dsz* promoter and associated regulatory regions of *Rhodococcus erythropolis* IGTS8. J. Bacteriol. 178: 6409-6418.

Lupas, A., Zühl, F., Tamura, T., Wolf, S., Nagy, I., De Mot, R., and Baumeister, W. 1997. Eubacterial proteasomes. Mol. Biol. Rep. 24: 125-131.

Mahillon, J., and Chandler, M. 1998. Insertion Sequences. Microbiol. Mol. Biol. Rev. 62: 725-774.

Masai, E., Sugiyama, K., Iwashita, N., Shimizu, S., Hauschild, J.E., Hatta, T., Kimbara, K., Yano, K., and Fukuda, M. 1997. The *bphDEF* meta-cleavage pathway genes involved in biphenyl/polychlorinated biphenyl degradation are located on a linear plasmid and separated from the initial *bphABC* genes in *Rhodococcus* sp. strain RHA1. Gene. 187: 141-149.

Masai, E., Yamada, A., Healey, J.M., Hatta, T., Kimbara, K., Fukuda, M., and Yano, K. 1995. Characterisation of biphenyl catabolic genes of Gram-positive polychlorinated biphenyl degrader *Rhodococcus* sp strain RHA1. Appl. Environ. Microbiol. 61: 2079-2085.

Meinhardt, F, Schaffrath, R., and Larsen, M. 1997. Microbial linear plasmids. Appl. Microbiol. Biotech. 47: 329-336.

Nagy, I., Schoofs, G., Compernolle, F., Proost, P., Vanderleyden, J. and De Mot, R., 1995b. Degradation of the thiocarbamate herbicide EPTC (*S*-ethyl dipropylcarbamothioate) and biosafening by *Rhodococcus* sp. NI86/21 involve an inducible cytochrome P-450 system and aldehyde dehydrogenase. J. Bacteriol. 177: 676-687.

Nagy, I., Schoofs, G., De Schrijver, A., Vanderleyden, J., and De Mot, R. 1997a. New method for RNA isolation from actinomycetes: application to the transcriptional analysis of the alcohol oxidoreductase gene *thcE* in *Rhodococcus* and *Mycobacterium*. Lett. Appl. Microbiol. 25: 75-79.

Nagy, I., Schoofs, G., Vanderleyden, J., and De Mot, R. 1997b. Transposition of the IS*21*-related element IS*1415* in *Rhodococcus erythropolis*. J. Bacteriol. 179: 4635-4638.

Nagy, I., Schoofs, G., Vanderleyden, J., and De Mot, R. 1997c. Further sequence analysis of the DNA regions with the *Rhodococcus* 20S proteasome structural genes reveals extensive homology with *Mycobacterium leprae*. DNA Seq. 7: 225-228.

Pelicic, V., Jackson, M., Reyrat, J-M., Jacobs, Jr.W.R., Gicquel, B., and Guilhot, C. 1997. Efficient allelic exchange and transposons mutagenesis in *Mycobacterium tuberculosis*. Proc. Natl. Acad. Sci. USA. 94: 10955-10960.

Picardeau, M., Bull, T.J., and Vincent, V. 1997. Identification and characterization of IS-like elements in *Mycobacterium gordonae*. FEMS Microbiol. Lett. 154: 95-102.

Picardeau, M., Varnerot, A., Rauzier, J., Gicquel, B. and Vincent, V. 1996. *Mycobacterium xenopi* IS*1395*, a novel insertion sequence expanding the IS*256* family. Microbiology. 142: 2453-2461.

Pisabarro, A., Correia, A., and Martin, J.F. 1998. Pulsed-field gel electrophoresis analysis of the genome of *Rhodococcus fascians*: genome size and linear and circular replicon composition in virulent and avirulent strains. Curr. Microbiol. 36: 302-308.

Poelarends, G.J., van Hylckama Vlieg, J.E.T., Marchesi, J.R., dos Santos, L.M.F., and Janssen, D.B. 1999. Degradation of 1,2-dibromoethane by *Mycobacterium* sp. strain GP1. J. Bacteriol. 181: 2050 – 2058.

Poelarends, G.J., Kulakov, L.A., Larkin, M.J., van Hylckama Vlieg, J.E.T., and Janssen, D.B. 2000. Roles of horizontal gene transfer and gene integration in evolution of 1,3-dichloropropene- and 1,2-dibromoethane-degradative pathways. J. Bacteriol. 182: 2191-2199.

Poelarends, G.J., Zandstra, M., Bosma, T., Kulakov, L.A., Larkin, M.J., Marchesi, J.R., Weightman, A.J., and Janssen, D.B. 2000. Haloalkane-Utilizing *Rhodococcus* strains isolated from geographically distinct locations possess a highly conserved gene cluster encoding haloalkane catabolism. J. Bacteriol. 182: 2725-2731.

Pogorelova, T.E., Ryabchenko, L.E., Sunzov, N.I., and Yanenko, A.S. 1996. Cobalt-dependent transcription of the nitrile hydratase gene in *Rhodococcus rhodochrous* M8. FEMS Microbiol. Lett. 144: 191-195.

Prauser, H. 1976. Host-phage relationships in nocardioform organisms. In: The Biology of the Nocardiae. Goodfellow, M., Brownell, G.H., and Serrano, J.A. eds. Academic Press, London. p 206-284.

Rådström, P., Sköld, O., Swedberg, G., Flensburg, J., Roy, P.H., and Sundström, L. 1994. Transposon Tn*5090* of plasmid R751, which carries an integron, is related to Tn*7*, Mu, and the retroelements. J. Bacteriol. 176: 3257-3268.

Rainey, F.A., Burghardt, J., Kroppenstedt, R.M., Klatte, S., and Stackebrandt, E. 1995. Phylogenetic analysis of the genera *Rhodococcus* and *Nocardia* and evidence for the evolutionary origin of the genus *Nocardia* from within the radiation of *Rhodococcus* species. Microbiology. 141: 523-528.

Redenbach, M., Scheel, J., and Schmidt, U. 2000. Chromosome topology and genome size of selected actinomycetes species. Antonie Van Leeuwenhoek International J. Gen. Mol. Microbiol. 78: 227-235.

Sankey, B.M., and Lau, P.C.K. 1998. Characterization of the basic replicon of *Rhodococcus* plasmid pSOX and development of a *Rhodococcus-Escherichia* coli shuttle vector. Appl. Env. Microbiol. 64: 4363-4367.

Schneider, D., Aigle, B., Leblond, P., Simonet, J.-M., and Decaris, B. 1993. Analysis of genome instability in *Streptomyces ambofaciens*. J. Gen. Microbiol. 139: 2559-2567.

Senda, T., Sugiyama, K., Narita, H., Yamamoto, T., Kimbara, K., Fukuda, M., Sato, M., Yano, K., and Mitsui, Y. 1996. Three-dimensional structures of free form and two substrate complexes of an extradiol ring-cleavage type dioxygenase, the BphC enzyme from *Pseudomonas sp*. Strain KKS102. J. Mol. Biol. 255: 735-752.

Shimizu, S., Kobayashi, H., Masai, E., and Fukuda, M. 2001. Characterization of the 450-kb linear plasmid in a polychlorinated biphenyl degrader, *Rhodococcus* sp. strain RHA1. Appl. Environ. Microbiol. 67: 2021-2028.

Shinohara, M., and Itoh, T. 1996. Specificity determinants in interaction of the initiator (Rep) proteins with the origins in the plasmids ColE2-P9 and ColE3-CA38 identified by chimera analysis. J. Mol. Biol. 257: 290-300.

Stark, W.M., Boocock, M.R., and Sherratt, D.J. 1992. Catalysis by site-specific recombinases. Trends Genet. 8: 432-439.

Steinmetz, M., Le Coq, D., Aymerich, S., Gonzy-Treboul, G., and Gay, P. 1985. The DNA sequence of the gene for the secreted *Bacillus subtilis* enzyme levansucrase and its genetic control sites. Mol. Gen. Genet. 200: 220-228.

Stolt, P., and Stoker, N.G. 1996a. Functional definition of regions necessary for replication and incompatibility in the *Mycobacterium fortuitum* plasmid pAL5000. Microbiology. 142: 2795-2802.

Stolt, P., and Stoker, N.G. 1996b. Protein-DNA interactions in the *ori* region of the *Mycobacterium fortuitum* plasmid pAL5000. J. Bacteriol. 178: 6693-6700.

Stolt, P., and Stoker, N.G. 1997. Mutational analysis of the regulatory region of the *Mycobacterium* plasmid pAL5000. Nucleic Acids Res. 25: 3840-3846.

Strohl, W.R. 1992. Compilation and analysis of DNA sequences associated with apparent streptomycete promoters. Nucleic Acids Res. 20: 961-974.

Takechi, S., Matsui, H., and Itoh, T. 1995. Primer RNA synthesis by plasmid-specified Rep protein for initiation of ColE2 DNA replication. EMBO Journal. 14: 5141-5147.

Tamura, T., Nagy, I., Lupas, A., Lottspeich, F., Cejka, Z., Schoofs, G., Tanaka, K., De Mot, R., and Baumeister, W. 1995. The first characterization of a eubacterial proteasome: the 20S complex of *Rhodococcus*. Curr. Biol. 5: 766-774.

Vereecke, D. 1997. Leafy gall induction by *Rhodococcus fascians*. PhD Thesis, University of Ghent, Belgium.

Vieira, J., and Messing, J. 1982. The pUC plasmids, an M13mp7-derived system for insertion mutagenesis and sequencing with synthetic universal primers. Gene. 19: 159-268.

Volff, J.-N., Vandewiele, D., Simonet, J-M., and Decaris, B. 1993. Stimulation of genetic instability in *Streptomyces ambofaciens* ATCC23877 by antibiotics that interact with DNA gyrase. J. Gen. Microbiol. 139: 2551-2558.

Wang, Y., Garnon, J., Labbé, D., Bergeron, H., and Lau, P.C.K. 1995. Sequence and expression of the *bpdC1C2BADE* genes involved in the initial steps of biphenyl/chlorobiphenyl degradation by *Rhodococcus* sp. M5. Gene. 164: 117-122.

Warhurst, A.M., and Fewson, C.A. 1994. Biotransformations catalyzed by the genus *Rhodococcus*. Critical Reviews In Biotechnology. 14: 29-73.

Yoon, J.H., Lee, J.S., Shin, Y.K., Park, Y.H., and Lee, S.T. 1997. Reclassification of *Nocaridioides simplex* ATCC 13260, ATCC 19565, and ATCC 19566 as *Rhodococcus erythropolis*. Int. J. Syst. Bacteriol. 47: 904-907.

Zheng, H., Thachuk-Saad, O., and Prescott, J.F. 1997. Development of a *Rhodococcus equi-Escherichia coli* plasmid shuttle vector. Plasmid. 38: 180-187.

From: *Genomics of GC-Rich Gram-Positive Bacteria*

3

Functional Genomics for the Discovery of New *Mycobacterium tuberculosis* Virulence Factors

Olivier Neyrolles and Brigitte Gicquel

Abstract

Eradication of tuberculosis remains one of the leading challenges for the next decades. The identification of *Mycobacterium tuberculosis* virulence factors might help to define new therapeutic as well as vaccine strategies. Recent advances in mycobacterial genetics and genomics are considerably promising to achieve such a goal.

Introduction

As defined by the last two G8 summits, one of the leading challenges for the next decades is the complete eradication of tuberculosis (TB), together with AIDS and malaria. Indeed, claiming 2-3 million lives each year, TB remains the most lethal infectious disease due to a single microbe, *i.e. Mycobacterium tuberculosis*. Reasons for such a dramatic situation include poor compliance of infected people to the intensive course of antibiotic therapy (6-9 months),

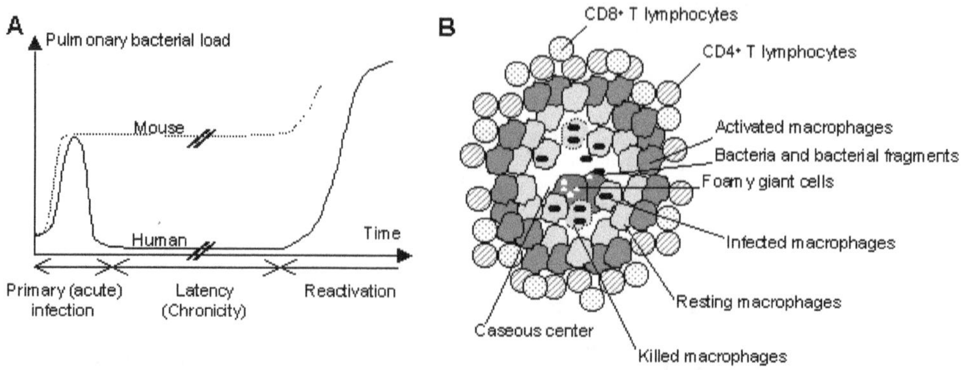

Figure 1. A. Evolution of *M. tuberculosis* infection in human (solid line) and mouse (dotted line). Note that latent phase in human is replaced by a chronic phase in the murine model, where mycobacterial load is sustained at a high level. B. Typical structure of a lung granuloma. Infected and lysed macrophages, together with giant cells in the center, are surrounded by activated macrophages and a mantle of lymphocytes. Upon reactivation, the caseous center liquifies and cavitates, leaving free bacteria in the airway for further dispersion in secondary foci, and causing a cough that leads to expelling free bacilli in the atmosphere.

the recent emergence of multi drug-resistant mycobacterial strains, together with deadly *M. tuberculosis*-HIV association and the relative inefficacy of the currently used vaccine, *Mycobacterium bovis* BCG, to prevent pulmonary TB, at least in adults. Therefore, a better understanding of the host-mycobacteria interactions is urgently needed. In particular, the virulence factors of *M. tuberculosis* remain mostly unknown and need to be identified. The knowledge of such factors could lead in a near future to the identification of targets for new anti-TB drugs, together with the development of new vaccines based on mycobacterial recombinant products and attenuated strains.

Microbial virulence is not simply an intrinsic microbial property, but rather a dynamic phenomenon including both the host and the microbe (Casadevall and Pirofski, 1999; Casadevall and Pirofski, 2001). In the case of *M. tuberculosis*, virulence mostly consists of persistence leading to immunopathology, rather than in "agressiveness" and cytopathogenicity (Honer zu Bentrup and Russell, 2001). Indeed, the human immune system generally copes well with *M. tuberculosis* primary infection (Figure 1A), and the efficacious cooperation between infected macrophages and T lymphocytes at the sites of infection leads to the destruction of most bacilli in typical structures called granulomas (Figure 1B) (Kaufmann, 2001). However, infection is rarely cleared in full, and some bacilli can persist inside the host creating a so-called "latent" infection that, even if asymptomatic and not transmissible, can reactivate in about 10% of the cases

(Figure 1A). Exacerbation of the host immune response upon reactivation (hypersensitivity) and collapse of the granulomatous lesions are responsible for chronic inflammation, deadly lung fibrosis and carcinoma, and dissemination of the bacilli to secondary foci. Latently infected people are currently thought to include up to one-third of the world's population, and are at high risk of developing fatal secondary TB. Furthermore, they are the most difficult to diagnose and to cure, and constitute a massive reservoir for further spreading of the bacilli.

Animal Models

Animal models to assess mycobacterial virulence mostly consist of guinea-pigs and mice. Guinea pigs are very sensitive to *M. tuberculosis* and develop a pathology that resembles necrotic lesions observed in humans, but the deeper understanding of the murine immune system and the availability of numerous murine knockout strains, makes the mouse model the most widely used. Following pulmonary inoculation of as few as 10^3 bacilli, mice, like humans, develop an acute phase of infection (Figure 1A) that can be used to monitore early manifestations of mycobacterial virulence, *i.e.* the ability to resist macrophage killing in the absence of a specific T cell response (Kaufmann, 2001). At this stage the mouse and human stories diverge since the infection is sustained at a high level and becomes chronic in mice only. Even if not identical to the latency period observed in humans, this chronic phase in the murine model is often used to investigate mycobacterial persistence and reactivation of mycobacterial infections. In other words, mutant strains of *M. tuberculosis* that are impaired in the establishment of a chronic phase of infection in mice are good candidates for beeing also impaired in persistence in humans.

Genetic Tools

For a long time, identification of mycobacterial virulence factors was severely hampered by the difficulty to cultivate pathogenic mycobacteria *in vitro* (*M. tuberculosis* doubling time ~20 h) and by the lack of genetic tools to manipulate mycobacterial genomes. The first mycobacterial virulence gene (the sigma factor RpoV-encoding gene) was discovered in 1995 only, by complementation experiments in a naturally attenuated strain of *M. bovis* (Collins *et al.*, 1995). Comparison of the *rpoV* sequences from the virulent and attenuated strains revealed a single point mutation leading to a Arg→His

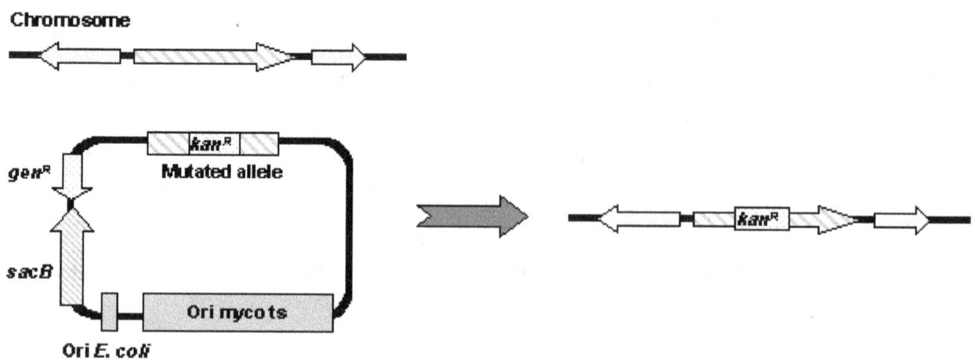

Figure 2. Schematic organisation of a plasmid allowing generation of allelic exchanges in *M. tuberculosis*. The construct can shuttle from *Escherichia coli* to mycobacteria and allows the delivery of a mutated allele containing a kanamycin-resistance cassette. Integrated backbone and circular plasmids are counter-selected by plating transformants on sucrose (lethal *Bacillus subtilis*-derived *sacB*) and at 39°C (thermo-sensitive mycobacterial Ori). This construct can also be used to generate transposon libraries. Adapted from Pelicic *et al.* (1997).

replacement at the C-terminus end of the protein, in a domain likely involved in DNA binding. Though the gene(s) regulated by RpoV have yet to be identified, it was the first report of a well-characterised mutation associated with an inability of the mutant mycobacterial strain to induce spleen lesions in guinea pigs.

In 1996-1997, genetic tools were developped to disrupt genes of interest by allelic replacement allowing the generation of marked or unmarked mutations (Pelicic *et al.*, 1996) (Figure 2). Such constructs gave a major boost to research in mycobacterial virulence and many groups have used these to inactivate their "home-genes" and study the phenotype of the deleted mutant strains, essentially in the mouse model. For instance, selective disruption of the *erp* gene encoding the surface protein Erp (*e*xported *r*epetitive *p*rotein) leads to a phenotype that is not able to establish an acute phase of infections in mice (Berthet *et al.*, 1998). Further *in vivo* and *in vitro* studies are now required to elucidate the precise function of Erp in *M. tuberculosis* early virulence.

Functional Genomics

The availability of the whole *M. tuberculosis* genome sequence in 1998 (Cole *et al.*, 1998) was regarded as an extraordinary opportunity to learn more about the virulence of Koch's bacillus and to give us the means to fight it:

"This history, together with the clues to conquer the tubercle bacillus, is written in its genome. (...) we now have the sequence of every potential drug target and of every antigen we may wish to include in a vaccine", said Douglas Young at the time (Young, 1998). Such enthousiasm was slightly dampened when *in silico* analysis of the whole genome failed to reveal the presence of any "obvious" virulence genes (a few phospholipase- and hemolysin-like-encoding genes have been detected but have yet to be proved to be involved in virulence), and showed that pathogenicity islands are missing in *M. tuberculosis*, whereas they are found in many other bacterial pathogens (Hacker and Carniel, 2001). Finally, a precise function has still to be assigned to more than half of the ~4,000 putative genes, and D. Young added : "But can we convert this mass of information into a usefull understanding?"

Functional genomics surely can!

Indeed, in spite of its apparent opaqueness, the whole *M. tuberculosis* genome sequence has definitely shed light on the search for mycobacterial virulence genes, and has paved the way for post-genomic research including comparative genomics, transcriptome and proteome analysis, as well as global and systematic genome analysis.

First of all, it gave researchers the opportunity to manipulate a gene of interest for which the sequence was unknown but the implication in virulence was suspected. One of the best examples of such an opportunity is given by the isocitrate lyase (ICL) story. ICL is a key-enzyme of the glyoxylate shunt that allows utilisation of fatty acids to generate carbohydrates in hypoxic conditions, a situation likely encountered in mycobacterial granulomas. An *in vitro* study of mycobacterial dormancy conducted in the early 1980's had reported that the glyoxylate shunt is preferentially used by bacteria in hypoxic conditions (Wayne and Lin, 1982). Almost twenty years later, an analysis of the *Mycobacterium avium*-infected macrophage proteome revealed that ICL is exclusive to intracellular bacteria (Honer Zu Bentrup *et al.*, 1999). Search for sequence similarities in the whole *M. tuberculosis* genome resulted in the identification of an *icl* gene, as well as in the design of the accurate strategy to inactivate it (McKinney *et al.*, 2000). The *icl*$^-$ strain of *M. tuberculosis* was found impaired in its ability to sustain a chronic phase of infection in the mouse model, revaling ICL as a major virulence factor likely to be involved in mycobacterial persistence.

Another example is given by the recent findings on mycobacterial nitrate reductase. It has been proposed that in the likely hypoxic milieu of granulomatous lesions, mycobacterial respiration might be coupled to the reduction of nitrates. Analysis of the *M. tuberculosis* H37Rv genome revealed the presence of four genes (*narG-J*) homologues of genes encoding nitrate reductases in other bacterial species such as *B. subtilis*. Cloning of the *M. bovis* BCG and *M. tuberculosis narG-J* clusters revealed that they encode a functional nitrate reductase *in vitro* (Weber *et al.*, 2000). Furthermore, a *narG*$^-$ strain of *M. bovis* BCG was found avirulent in a SCID mouse model, suggesting that *narG-J* is a virulence gene cluster in *M. tuberculosis* (Fritz *et al.*, 2002).

A major limitation to this type of strategy is that it can only be used for genes with a predicted function. As we mentionned above, more than half of the 4,000 *M. tuberculosis* putative genes are of unkown, if any, function. Since it is not practical to knock-out all these genes, global approaches are necessary to evaluate their putative implication in mycobacterial virulence.

Comparative genomics between virulent and attenuated strains was supposed to be an attractive and elegant way to uncover putative virulence genes in mycobacteria. Sixteen *M. tuberculosis* DNA fragments (RD1-16) have been shown to be missing in the genome of the vaccine attenuated strain *M. bovis* BCG as compared to *M. tuberculosis* (Behr *et al.*, 1999; Gordon *et al.*, 1999; Mahairas *et al.*, 1996). Most of these fragments, except RD1-3, are also present in the virulent parental species *M. bovis*, indicating that they are unlikely to bear major virulence genes. RD1 is the only *M. tuberculosis* DNA fragment missing in all the *M. bovis* BCG strains tested, and might account for the original attenuation of BCG noted by Calmette and Guérin in the late 1920's. This fragment encompasses genes coding for the *M. tuberculosis*-restricted ESAT-6 family members, as well as genes encoding members of the PE and PPE family. PE and PPE proteins are long polypeptides containing Pro-Glu and Pro-Pro-Glu repeated motifs, respectively (Cole *et al.*, 1998). Their function is still unkown but they have proposed to be implicated in antigenic variations and inhibition of the MHC-I processing pathway. These features have yet to be tested but if proved, this could make these proteins major virulence factors of *M. tuberculosis*.

Analysis of the *M. tuberculosis* transcriptome revealed a set of a few genes that are expressed in human macrophages only, as compared to *in vitro* grown mycobacteria (Graham and Clark-Curtiss, 1999). Among these genes is the *icl*, confirming the relevance of the strategy. Other genes are reported in Table 1 and now need to be tested using inactivation strategies.

Table 1. *M. tuberculosis* genes selectively induced inside human macrophages. These genes were revealed by transcriptome analysis of bacteria recovered from 48-h infected cells. Adapted from Graham and Clark-Curtiss (1999).

Gene Name	Putative function
sigE	Alternative sigma factor
sigH	Alternative sigma factor
aceA	Isocitrate lyase
ponA	Penicillin-binding
pks2	Polyketide synthase
uvrA	Endonuclease
ctpV	Cation transporter
Rv0170	*mce1* operon
Rv3070	Unknown
Rv3483c	Unknown
Rv0903c	Two-component transcriptional activator

Proteome analysis has been successfully used in *M. tuberculosis* (Jungblut *et al.*, 2001; Jungblut *et al.*, 1999) and a recent study revealed the presence of twenty proteins in *M. tuberculosis* that are missing in *M. bovis* BCG and were not reported in the sequences of the RD1-16 DNA fragments (Mattow *et al.*, 2001). These genes mostly encode enzymes and transporters, and now need to be studied one by one. Proteome analysis of intracellular bacteria as well as of bacilli recovered from infected animals should lead to interesting results in the future.

Finally, the knowledge of the whole genome has greatly facilitated global approaches in functional genomics.

Generation of mutant libraries of *M. tuberculosis* by random transposition are now possible using the same construct as the one described in Figure 2, and efficient strategies have been set up to screen ordered transposon-mediated mutant libraries on 96-well microplates (Jackson *et al.*, 1999). This has allowed the detection of virulence genes by signature-tagged mutagenesis (STM), a method first described in 1995 in a *Salmonella* model (Hensel *et al.*, 1995). The principle of STM is to screen large libraries of individual strains, each of them containing a unique DNA tag at a particular location in the genome. Tags from a mixed population of *M. tuberculosis* mutants representing the inoculum and mycobacteria recovered from the lungs of infected mice were detected and compared using PCR and

Table 2. *M. tuberculosis* virulence genes identified by STM. Adapted from Camacho *et al.* (1999).

Gene name	Putative function
fadD26	Acyl-CoA synthase
pksb	Polyketide synthase
lipF	Lipid degradation
mmpL7	Proton-dependent transporter
drrC	ABC transporter
mmpL2	Proton-dependent transporter
modA	Molybdenum fixation
mmpL4	Proton-dependent transporter
Rv2452c	Unknown
Rv1395	Transcriptional activator
Rv0204	Unknown
Rv3018c	PPE family

hybridisation. Screening of a 2,000-member mutant library allowed the identification of 12 genes implicated in mycobacterial virulence (Table 2).

Three of these genes were found clustered in a 50 kb DNA region involved in the synthesis of phtiocerol dimycoserosates (DIMs, Figure 3) (Camacho *et al.*, 2001; Camacho *et al.*, 1999). In particular, inactivation of the *fadD26*, *drrc* and *mmpL7* genes were found to impair synthesis or transport of the DIMs and to lead to attenuated virulence.

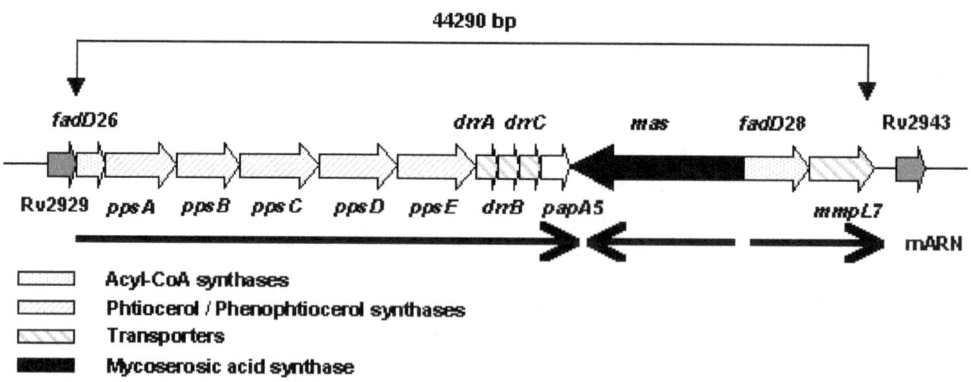

Figure 3. Organisation of the gene cluster involved in synthesis and transport of phtiocerol dimycoserosates in *M. tuberculosis* (Camacho *et al.*, 1999). Arrows underline co-transcribed genes (operons).

New *in vivo* as well as *in vitro* screening methods are now underway in the search for genes implicated in mycobacterial persistence and intracellular replication. In particular STM libraries can be screened at the sub-cellular organelle level, in order to identify genes involved in the unique ability of pathogenic mycobacteria to arrest the maturation of their phagosome when inside macrophages (Russell, 2001). Such genes are likely to be major virulence genes in *M. tuberculosis* and have yet to be identified.

Acknowledgements

We wish to thank Thomas Roe (Imperial College, London) and Mary Jackson (Institut Pasteur, Paris) for carefully reading the manuscript and helpfull discussions.

References

Behr, M.A., Wilson, M.A., Gill, W.P., Salamon, H., Schoolnik, G.K., Rane, S. and Small, P.M. 1999. Comparative genomics of BCG vaccines by whole-genome DNA microarray. Science 284: 1520-1523.

Berthet, F.X., Lagranderie, M., Gounon, P., Laurent-Winter, C., Ensergueix, D., Chavarot, P., Thouron, F., Maranghi, E., Pelicic, V., Portnoi, D., Marchal, G. and Gicquel, B. 1998. Attenuation of virulence by disruption of the *Mycobacterium tuberculosis erp* gene. Science 282: 759-762.

Camacho, L.R., Constant, P., Raynaud, C., Laneelle, M.A., Triccas, J.A., Gicquel, B., Daffe, M. and Guilhot, C. 2001. Analysis of the phthiocerol dimycocerosate locus of *Mycobacterium tuberculosis*. Evidence that this lipid is involved in the cell wall permeability barrier. J. Biol. Chem. 276: 19845-19854.

Camacho, L.R., Ensergueix, D., Perez, E., Gicquel, B. and Guilhot, C. 1999. Identification of a virulence gene cluster of *Mycobacterium tuberculosis* by signature-tagged transposon mutagenesis. Mol. Microbiol. 34: 257-267.

Casadevall, A. and Pirofski, L. 2001. Host-pathogen interactions: the attributes of virulence. J. Infect. Dis. 184: 337-344.

Casadevall, A. and Pirofski, L.A. 1999. Host-pathogen interactions: redefining the basic concepts of virulence and pathogenicity. Infect. Immun. 67: 3703-3713.

Cole, S.T., Brosch, R., Parkhill, J., Garnier, T., Churcher, C., Harris, D., Gordon, S.V., Eiglmeier, K., Gas, S., Barry, C.E., 3rd, Tekaia, F., Badcock, K., Basham, D., Brown, D., Chillingworth, T., Connor, R., Davies, R.,

Devlin, K., Feltwell, T., Gentles, S., Hamlin, N., Holroyd, S., Hornsby, T., Jagels, K., Barrell, B.G. and *et al*. 1998. Deciphering the biology of *Mycobacterium tuberculosis* from the complete genome sequence. Nature 393: 537-544.

Collins, D.M., Kawakami, R.P., de Lisle, G.W., Pascopella, L., Bloom, B.R. and Jacobs, W.R., Jr. 1995. Mutation of the principal sigma factor causes loss of virulence in a strain of the *Mycobacterium tuberculosis* complex. Proc. Natl. Acad. Sci. U. S. A. 92: 8036-8040.

Fritz, C., Maass, S., Kreft, A. and Bange, F.C. 2002. Dependence of *Mycobacterium bovis* BCG on anaerobic nitrate reductase for persistence is tissue specific. Infect. Immun. 70: 286-291.

Gordon, S.V., Brosch, R., Billault, A., Garnier, T., Eiglmeier, K. and Cole, S.T. 1999. Identification of variable regions in the genomes of tubercle bacilli using bacterial artificial chromosome arrays. Mol. Microbiol. 32: 643-655.

Graham, J.E. and Clark-Curtiss, J.E. 1999. Identification of *Mycobacterium tuberculosis* RNAs synthesized in response to phagocytosis by human macrophages by selective capture of transcribed sequences (SCOTS). Proc. Natl. Acad. Sci. U. S. A. 96: 11554-11559.

Hacker, J. and Carniel, E. 2001. Ecological fitness, genomic islands and bacterial pathogenicity. A Darwinian view of the evolution of microbes. EMBO Rep. 2: 376-381.

Hensel, M., Shea, J.E., Gleeson, C., Jones, M.D., Dalton, E. and Holden, D.W. 1995. Simultaneous identification of bacterial virulence genes by negative selection. Science 269: 400-403.

Honer Zu Bentrup, K., Miczak, A., Swenson, D.L. and Russell, D.G. 1999. Characterization of activity and expression of isocitrate lyase in Mycobacterium avium and *Mycobacterium tuberculosis*. J. Bacteriol. 181: 7161-7167.

Honer zu Bentrup, K. and Russell, D.G. 2001. Mycobacterial persistence: adaptation to a changing environment. Trends Microbiol. 9: 597-605.

Jackson, M., Raynaud, C., Laneelle, M.A., Guilhot, C., Laurent-Winter, C., Ensergueix, D., Gicquel, B. and Daffe, M. 1999. Inactivation of the antigen 85C gene profoundly affects the mycolate content and alters the permeability of the *Mycobacterium tuberculosis* cell envelope. Mol. Microbiol. 31: 1573-1587.

Jungblut, P.R., Muller, E.C., Mattow, J. and Kaufmann, S.H. 2001. Proteomics reveals open reading frames in *Mycobacterium tuberculosis* H37Rv not predicted by genomics. Infect. Immun. 69: 5905-5907.

Jungblut, P.R., Schaible, U.E., Mollenkopf, H.J., Zimny-Arndt, U., Raupach, B., Mattow, J., Halada, P., Lamer, S., Hagens, K. and Kaufmann, S.H. 1999.

Comparative proteome analysis of *Mycobacterium tuberculosis* and *Mycobacterium bovis* BCG strains: towards functional genomics of microbial pathogens. Mol. Microbiol. 33: 1103-1117.

Kaufmann, S.H.E. 2001. How can immunology contribute to the control of tuberculosis? Nat. Rev. Immunol. 1: 20-30.

Mahairas, G.G., Sabo, P.J., Hickey, M.J., Singh, D.C. and Stover, C.K. 1996. Molecular analysis of genetic differences between *Mycobacterium bovis* BCG and virulent M. bovis. J. Bacteriol. 178: 1274-1282.

Mattow, J., Jungblut, P.R., Schaible, U.E., Mollenkopf, H.J., Lamer, S., Zimny-Arndt, U., Hagens, K., Muller, E.C. and Kaufmann, S.H. 2001. Identification of proteins from *Mycobacterium tuberculosis* missing in attenuated *Mycobacterium bovis* BCG strains. Electrophoresis 22: 2936-2946.

McKinney, J.D., Honer zu Bentrup, K., Munoz-Elias, E.J., Miczak, A., Chen, B., Chan, W.T., Swenson, D., Sacchettini, J.C., Jacobs, W.R., Jr. and Russell, D.G. 2000. Persistence of *Mycobacterium tuberculosis* in macrophages and mice requires the glyoxylate shunt enzyme isocitrate lyase. Nature 406: 735-738.

Pelicic, V., Jackson, M., Reyrat, J.M., Jacobs, W.R., Jr., Gicquel, B. and Guilhot, C. 1997. Efficient allelic exchange and transposon mutagenesis in *Mycobacterium tuberculosis*. Proc. Natl. Acad. Sci. U. S. A. 94: 10955-10960.

Pelicic, V., Reyrat, J.M. and Gicquel, B. 1996. Generation of unmarked directed mutations in mycobacteria, using sucrose counter-selectable suicide vectors. Mol. Microbiol. 20: 919-925.

Russell, D.G. 2001. *Mycobacterium tuberculosis*: here today, and here tomorrow. Nat. Rev. Mol. Cell Biol. 2: 569-577.

Wayne, L.G. and Lin, K.Y. 1982. Glyoxylate metabolism and adaptation of *Mycobacterium tuberculosis* to survival under anaerobic conditions. Infect. Immun. 37: 1042-1049.

Weber, I., Fritz, C., Ruttkowski, S., Kreft, A. and Bange, F.C. 2000. Anaerobic nitrate reductase (narGHJI) activity of *Mycobacterium bovis* BCG in vitro and its contribution to virulence in immunodeficient mice. Mol. Microbiol. 35: 1017-1025.

Young, D.B. 1998. Blueprint for the white plague. Nature 393: 515-516.

From: *Genomics of GC-Rich Gram-Positive Bacteria*

4

Global Analysis of *Streptomyces*

Jean-Luc Pernodet and Philippe Mazodier

Abstract

Streptomyces spp. are gram-positive, filamentous, soil bacteria. They belong to the order Actinomycetales. The Streptomycetes are remarkable in terms of the morphological and metabolic differentiation phenomena that they present. The first step in development is the germination of a spore, giving rise to a basal mycelium, which develops in close contact with the nutritive substrate; the basal mycelium gives rise to aerial hyphae, which ramify and differentiate into spores. This morphological differentiation is accompanied by metabolic differentiation. Indeed, during the later stages of this development *Streptomyces* generally synthesise a large number of very diverse secondary metabolites, the best known of which are antibiotics. *Streptomyces* and closely related genera produce more than 50% of the 200 commercially available antibiotics. The third remarkable feature of *Streptomyces* spp. is the structure of their genome. They possess a large chromosome, which consists an ~8 Mb linear DNA molecule. The first genome sequence of a *Streptomyces* was completed in July 2001. The aim of this chapter is to present briefly the genomic data available for this group of bacteria and to examine how global approaches can be used in the domains mentioned above: genome architecture, differentiation and secondary metabolism.

Introduction

Streptomyces spp. are gram-positive, filamentous, soil bacteria. They belong to the order Actinomycetales and are characterised by the high G+C ratio of their DNA. Streptomycetes are remarkable in terms of the morphological and metabolic differentiation phenomena that they present (Hopwood, 1999). The first step in the development of *Streptomyces* is the germination of a spore, giving rise to a basal or substrate mycelium, which develops in close contact with the nutritive substrate. In response to poorly identified signals emanating from cellular metabolism and from the environment, the basal mycelium gives rise to aerial hyphae, which ramify and differentiate into spores (Chater, 1998; Chater, 2001). During their complex life cycle, *Streptomyces* spp. differentiate into at least three distinct cell types: substrate hyphae, aerial hyphae and spores. The passage from one mycelial state to the other is marked by a pause in growth and requires the expression of numerous genes specific to the aerial mycelium.

This morphological differentiation is accompanied by metabolic differentiation. Indeed, during the later stages of this development *Streptomyces* spp. generally synthesise a large number of very diverse secondary metabolites, the best known of which are antibiotics. *Streptomyces* and closely related genera produce more than 50% of the 200 commercially available antibiotics and most of the 10,000 known antibiotics. Secondary metabolism also produces bioactive compounds that are used as antifungal, anti-parasitic, antiviral, antitumoural or immunosuppressant agents or as insecticides or herbicides. *Streptomyces* spp. are also able to synthesise a large number of degradative enzymes, some of which are produced industrially.

The third remarkable feature of *Streptomyces* is the structure of their genome. They possess a large chromosome, which consists of an ~8 Mb linear DNA molecule (Lin *et al.*, 1993; Volff and Altenbuchner, 2000). Several unique accessory genetic elements, such as large linear plasmids (Kinashi *et al.*, 1987; Kinashi *et al.*, 1992) or integrative elements (Bibb *et al.*, 1981; Boccard *et al.*, 1989; Hopwood *et al.*, 1984), have been discovered in this group of bacteria.

Due to the important role of *Streptomyces* products in medicine, agronomy and the food industry, these bacteria have been studied in detail by academic and industrial laboratories. The aim of this chapter is to present briefly the genomic data available for this group of bacteria and to examine how global approaches can be used in the domains mentioned above: genome architecture, differentiation and secondary metabolism. As the *Streptomyces coelicolor* genome sequence was completed only recently (July 2001), there are few publications concerning this topic although many projects have been undertaken.

Data Available

Strains Studied and Genomes Sequenced

David Hopwood recently reviewed the development of *Streptomyces* genetics (Hopwood, 1999). By far the best known *Streptomyces* is *S. coelicolor* A3(2). Four decades of studies have provided a comprehensive picture of this organism. A large part of this work, pioneered by David Hopwood, was carried out at the John Innes Centre (Norwich, UK; http://www.jic.bbsrc.ac.uk/corporate/Science_Departments/mol_micro.html). *S. coelicolor* is the model strain studied in many academics laboratories working on *Streptomyces* worldwide, as strains and the genetic tools developed for this microorganism are readily available. Most of the great findings specific to *Streptomyces* were discovered or documented in this strain. In 1996, an ordered cosmid library and a detailed genetic and physical map of *S. coelicolor* were established. A set of 319 cosmids covered nearly the whole chromosome (Redenbach *et al.*, 1996). A collaboration between the Sanger Centre (now Institute) and the John Innes Centre successfully sequenced the entire *S. coelicolor* genome (http://www.sanger.ac.uk/Projects/S_coelicolor/). This project was funded by the BBSRC (Biotechnology and Biological Sciences Research Council) and the Beowulf Genomics initiative of the Wellcome Trust. The work began in August 1997 and the sequence was generated with a clone-by-clone approach, initially using the 319 ordered cosmids and subsequently using a BAC library to fill in the gaps and to confirm the assembly. The sequence was completed in July 2001. During the entire sequencing period the sequence was accessible on the Sanger Centre website. This site was updated on a daily basis and Blast searches could be performed. This access to the sequence data was precious for those working on *Streptomyces*.

Streptomyces lividans 66 is closely related to *S. coelicolor* A3(2) (Hopwood *et al.*, 1983). The two strains were used and studied in parallel in many laboratories. The main reason for this is that during the 1980s it was much easier to use *S. lividans* than *S. coelicolor* as a host for DNA cloning. It was subsequently shown that methylated DNA, such as that extracted from *Escherichia coli* strains, cannot be introduced into *S. coelicolor* (MacNeil *et al.*,1992; Flett *et al.*, 1997). Chromosome instability was studied in great detail in *S. lividans* and *Streptomyces ambofaciens* (Fischer *et al.*, 1997; Redenbach *et al.*, 1993; Volff and Altenbuchner, 1998). The current availability of many *S. lividans* mutants perpetuates the interest in this microorganism. As *S. lividans* and *S. coelicolor* are very closely related, most of the sequence data obtained for *S. coelicolor* can be used to study genes from *S. lividans*.

S. coelicolor and *S. lividans* produce several antibiotics, none of which are of commercial value. This absence of commercial issues favoured the exchange of strains and information within the academic community.

Many other different strains, belonging to several *Streptomyces* species, have been studied to various extents but most *Streptomyces* strains are only considered and exploited for one aspect of their life style, the production of a given antibiotic. *Streptomyces avermitilis* produces several secondary metabolites, one of which, avermectin, is widely used as an antiparasitic agent in human and veterinary medicine. The sequence of *S. avermitilis* MA-4680 was determined in Japan through a collaboration involving several academic research institutions (http://www.bio.nite.go.jp/coll/Strept_e.htm). However, the complete sequence is not yet publicly available. A whole genome, shotgun sequencing strategy was employed to sequence *S. avermitilis* (Omura *et al.*, 2001).

Two other strains of industrial interest have been sequenced by private companies. *Saccharopolyspora spinosa*, although not a Streptomycete, belongs to a related Actinomycete genus and produces spinosin, commercialised for its insecticide action (Waldron *et al.*, 2001). In March 1999, Dow AgroSciences LLC and Integrated Genomics Inc. announced a research and licensing agreement for the sequencing of *Saccharopolyspora spinosa*. Diversa Corporation isolated *Streptomyces diversa*™, which is used as a host for gene expression and to study pathways. In February 2001, Diversa Corporation and Celera Genomics announced the completion of the entire genome sequence of *S. diversa*™. The sequences of *S. spinosa* and *S. diversa*™ are not publicly available.

The sequences of the ends of the *Streptomyces ambofaciens* chromosome, which produces spiramycin, will be determined in France by Genoscope, the French national sequencing centre, as part of a collaboration with two academic laboratories (http://www.genoscope.cns.fr/externe/English/Projets/Projet_DL/DL.html).

Genes from Streptomyces coelicolor

The genome of *S. coelicolor* A3(2) is 8,667,507 base pairs long, which is about twice the size of that of the model bacteria *Escherichia coli* and *Bacillus subtilis*. At total of 7825 putatively coding sequences (CDSs) have been identified, together with 55 pseudo-genes, 63 tRNA genes and 6 rRNA operons. Therefore, *S. coelicolor* possesses more genes than the model yeast, *Saccharomyces cerevisiae*, which contains about 6000 genes, and about a quarter of the estimated number of human genes (31,000). Only 6% of these 7825 CDSs have been studied previously; 41% are similar to genes of known function, 30% are similar to CDSs of unknown function and 23% are novel (Bentley, 2001).

The sequence shows many genes (65) encoding potential sigma factors, most (51) of which are of the ECF (Extra Cytoplasm Factor) type (Lonetto *et al.*, 1994). A large number of sigma factors was to be expected given the notorious heterogeneity of the *Streptomyces* promoter regions (Bourn and Babb, 1995; Strohl, 1992). Indeed, bacterial sigma factors are responsible for the binding of the RNA polymerase to the promoter and confer sequence specificity. However, the actual number – nearly three times as many as in the sequenced genomes of some other soil bacteria – was a surprise.

As expected for a soil organism living in a rapidly changing environment, *S. coelicolor* contains a large number of genes encoding enzymes involved in the degradation of substrates (e.g. agarase, cellulase, xylanase) and there are 81 typical ABC permeases and 141 ATP-binding proteins which might be involved in the influx and efflux of metabolites.

Not surprisingly, *S. coelicolor* also contains numerous regulators of various types (42 LysR-like, 24 AraC-like, 53 GntR-like and 125 TetR-like regulators), which are involved in the growth, survival and reproduction of this highly sophisticated and organised microorganism.

However, some unexpected genes are also present in *S. coelicolor,* suggesting that we should reassess some of the fundamental ways we consider *Streptomyces*. In particular, *S. coelicolor* contains two sets of genes that are orthologous to the genes involved in the formation of gaseous vacuoles in

Cyanobacteria. Although this type of structure has never been reported in *Streptomyces*, several *Streptomyces* strains have recently been found in a marine environment where the formation of gaseous vacuoles may be relevant.

The presence of three clusters of genes orthologous to *narGHIJ* encoding subunits of respiratory nitrate reductase and involved in anaerobic growth of *B. subtilis* on nitrate, is very puzzling because *Streptomyces* are believed to be exclusively aerobic bacteria. This observation should prompt us to study the survival of *Streptomyces* in anaerobic or microaerobic conditions.

The presence of many cryptic or silent secondary metabolic pathways was also surprising. Indeed, *S. coelicolor* was known to possess four antibiotic pathways. Three are chromosomally encoded, actinorhodin, undecylprodigiosin, and the calcium dependent antibiotic (CDA), and one (methylenomycin) is encoded by a large linear plasmid (SCP1). Sequencing revealed many putative secondary metabolic pathways involved in the production of geosmin, sesquiterpene, chalcone and peptide siderophores, such as coelichelin (Challis and Ravel, 2000). The genes encoding these biosynthetic pathways account for around 5% of the total chromosome.

Sequencing confirmed the presence of numerous duplicated genes, a situation thought to be related to the diphasic life cycle of *S. coelicolor*. This was shown to be the case for the two copies of the genes involved in glycogen formation and utilisation. One of the copies is expressed in the basal mycelium and the other in the aerial mycelium (Schneider *et al.*, 2000).

The functions of well over half of the proteins encoded by the *S. coelicolor* are not yet known. Since even putting a gene into the 'similar to known' class often does not tell us much about the biological role, eg. a gene is likely to encode a phosphatase, but what is the substrate?.The elucidation of the role of these genes is one of the aspects of the project entitled "*Making the most of a genome sequence: the application of global transcriptome and proteome analysis of Streptomyces coelicolor A3(2)*" which is funded by the BBSRC (Grant No: FGT11408).

Structure of *Streptomyces* Genomes

Genetic and Physical Maps

S. coelicolor A3(2) was the first *Streptomyces* species for which a genetic map was established through extensive conjugative mating experiments involving differentially marked strains (Hopwood, 1965). This genetic map is circular, as confirmed by the analysis of recombinants obtained by protoplast fusion (Hopwood and Wright, 1978; Hopwood *et al.*, 1977). This characteristic was extended to other *Streptomyces* species. The genetic map of *S. coelicolor* contains the most markers, but these markers are not randomly distributed: two silent regions, almost devoid of markers, were located in two opposite quadrants. One of these later became less apparent as more genes were mapped, but the other remained striking and particularly large (covering almost one quarter of the entire genome).

Pulse-field gel electrophoresis was used to determine that the size of the *Streptomyces* chromosome was about 8 Mb (Kieser *et al.*, 1992). Macro-restriction maps and the demonstration that free ends carried covalently bound proteins proved that the chromosomal DNA was linear in *S. lividans* and *S. coelicolor* (Lin *et al.*, 1993). This linearity, which as been confirmed for other *Streptomyces* species (Leblond *et al.*, 1996; Lezhava *et al.*, 1995) and for related Actinomycetes such as *Saccharopolyspora erythraea* (Reeves *et al.*, 1998), seems to be a characteristic of this group of bacteria. However, it is not encountered in all Actinomycetes, because *Mycobacterium* and *Corynebacterium* have circular chromosomes (Redenbach *et al.*, 2000). The physical map of *S. coelicolor* confirmed that the large silent region was physically long and encompassed the two extremities of the linear chromosome. The linearity of the chromosome is not absolutely required in *Streptomyces*: circular chromosomes can be generated by targeted recombination to join the two ends together (Lin *et al.*, 1993). Indeed, circular chromosomes can arise spontaneously as the result of genetic instability (Volff and Altenbuchner, 2000; Volff *et al.*, 1997).

How can a circular genetic map be obtained from a linear genome? Stahl and Steinberg proposed one explanation (Stahl, 1967; Stahl and Steinberg, 1964). If a linear chromosome recombines with a strong bias towards an even number of crossovers, this creates genetic linkage between markers at opposite ends of the chromosome. This type of bias was observed during the analysis of telomeres in recombinant chromosomes obtained after plasmid-mediated interspecific conjugation between *S. coelicolor* and *S. lividans*

(Wang *et al.*, 1999). This bias could be explained by the merozygosity of the recombining chromosome and/or by the physical association of the termini, probably due to the interaction between covalently bound terminal proteins. Cytological evidence for the association of the ends of the linear *S. coelicolor* chromosome was recently obtained by fluorescence *in situ* hybridisation (Yang and Losick, 2001). In several *Streptomyces* species the ends of the linear chromosome contain large (25 to 550 kb) terminal inverted repeats (TIR) (Volff and Altenbuchner, 2000). However, these long TIR are absent from *S. avermitilis*, in which a 174-bp sequence was the only terminal repeat sequence present (Omura *et al.*, 2001).

Replication of Linear Replicons

The first linear bacterial plasmid was isolated from *Streptomyces* (Hayakawa *et al.*, 1979a; Hayakawa *et al.*, 1979b). Large linear plasmids, of up to several hundred kb, were later discovered in *Streptomyces* (Kinashi *et al.*, 1987). The replication of these linear replicons (plasmid and chromosome) has received a lot of attention recently. In all cases studied so far, there is a centrally located origin of replication from which bi-directional replication starts (Chang and Cohen, 1994; Chang *et al.*, 1996; Musialowski *et al.*, 1994; Redenbach *et al.*, 1999). This produces a 3' overhang of the leading strand at the ends of the chromosome and patching is required to obtain blunt-ended DNA molecules (Chang and Cohen, 1994). The precise nature of the patching mechanism is still unknown (Chen, 1996; Qin and Cohen, 1998). The termini of several *Streptomyces* linear replicons have been sequenced, revealing that 166-168 bp at the extremities are highly conserved and can form complex secondary structures (Huang *et al.*, 1998). *Streptomyces rochei* and *S. lividans* genes encoding terminal proteins were recently isolated and characterised, and an *S. coelicolor* homologue has been identified in the sequenced genome (Bao and Cohen, 2001). The *S. coelicolor* terminal protein gene is located about 120 kb from the end of the *S. coelicolor* chromosome. In *S. lividans,* inactivation of the terminal protein gene, *tpgL,* prevents the propagation of cells that contain a linear chromosome, but not that of cells containing a circular chromosome. Similarly, the inactivation of *tpgL* also prevents the propagation of cells containing a linear plasmid but not carrying a gene encoding for a terminal protein. This indicates that these terminal proteins have a specific and essential role in the replication of linear DNA.

Gene Distribution

One of the most original features of the *S. coelicolor* chromosome is the distribution of the genes. In the linear chromosome, all of the essential genes are located in the central region rather than at the extremities. This was clearly shown by the physical and genetic maps of *S. coelicolor* deduced from the ordered cosmid library (Redenbach *et al.*, 1996). The only gene involved in macromolecular synthesis or primary metabolism that has been found in the terminal regions is *argG,* which is involved in arginine biosynthesis. The sequence data confirms this observation: essential genes are almost completely absent from the first 1500 kb of the left end and from the first 2700 kb of the right end. This does not mean that these regions are devoid of genes - the gene density of these terminal regions is similar to that of the central part of the chromosome - but the genes found in these regions are non-essential or conditionally essential. Many transposons are also present in the terminal regions (Bentley, 2001) (http://www.sanger.ac.uk/Projects/S_coelicolor/).

The completion of the genome sequences of *S. coelicolor* and of other Actinomycetes, such as *Mycobacterium tuberculosis* (4.4 Mb circular chromosome) (Cole *et al.*, 1998) and *Corynebacterium diptheriae* (2.5 Mb circular chromosome) (http://www.sanger.ac.uk/Projects/C_diphtheriae/), allowed whole genome comparisons. Genes homologous to genes from other Actinomycetes were found in the central part of the *S. coelicolor* genome. Many inversions, symmetrical around the origin, are postulated to have occurred in these organisms. It has been suggested that the *Streptomyces* chromosome evolved from a smaller, ancestral circular chromosome that became linear (Bentley, 2001; Volff and Altenbuchner, 2000). The size of this ancestral chromosome then increased as a result of gene duplications or acquisitions by horizontal transfer, particularly in the terminal part of the chromosome.

The concentration of essential genes in the central region seems to be a common feature in *Streptomyces*, as shown by the viability of strains of many species with large deletions in the terminal parts (see below). When other *Streptomyces* sequences become available, comparisons will confirm whether the central part is highly conserved and the extremities are much more variable, both in terms of organisation and the nature of the genes present.

Genetic Instability

The *Streptomyces* chromosome is relatively unstable (Volff and Altenbuchner, 2000). This phenomenon has been studied mostly in *S. lividans* (Altenbuchner and Cullum, 1984; Flett and Cullum, 1987; Rauland *et al.*, 1995) and *S. ambofaciens* (Dary *et al.*, 2000; Leblond *et al.*, 1990; Simonet *et al.*, 1992). Large deletions, sometimes accompanied by DNA amplifications, occur spontaneously in more than 0.1% of germinating spores. These deletions can result in the loss of up to 2 Mb of chromosomal DNA. Most of deletions occur at or near the chromosomal ends and may lead to circularisation of the chromosome. The strains with these large terminal deletions are still viable in laboratory conditions, but they often have altered antibiotic biosynthesis and resistance, morphological differentiation and synthesis of degradative enzymes. The mechanisms leading to these deletions are not fully understood, but it is clear that instability is not due to linearity, because spontaneously or artificially circularised *Streptomyces* chromosomes are as unstable as (or sometimes more unstable than) linear ones (Volff *et al.*, 1997). This instability could be due to the collapse of replication forks at single strand breaks (Volff and Altenbuchner, 1998). *Streptomyces* spp. might not have a higher degree of intrinsic genetic instability than other bacteria; instead they might be better able to tolerate large deletions, which would be lethal in other bacteria. This high level of tolerance is due to the architecture of their chromosome (Volff and Altenbuchner, 2000).

The extremities of the chromosome are also affected by other type of rearrangements, such as the exchange of ends between chromosomes and linear plasmids or the exchange of ends between the two arms of a chromosome (Fischer *et al.*, 1998; Gravius *et al.*, 1994). These exchanges could lead to gene duplication and genetic redundancy. The terminal regions of *Streptomyces* chromosomes might show a higher degree of genetic plasticity than the central region. This could lead to the more rapid evolution of these terminal regions and of the genes that they harbour.

Morphological Differentiation.

Bald Mutants

Two classes of differentiation mutants have been identified; "*bald*" mutants are unable to make aerial mycelium and "*white*" mutants are unable to sporulate (Chater, 1998).

Bald mutants can be divided into complementation groups, which can be organised in a cascade according to extracellular complementation (Willey *et al.*, 1993) Indeed, when certain pairs of *bld* mutants are grown in close vicinity, the ability to form aerial mycelium can be restored in one of the strains. As the relationship is unidirectional, examining all the possible pairs of bald mutants led to the construction of a hierarchy of genes based on the ability of each *bld* mutant to restore the ability to form aerial mycelium to the others, as follows: *bldJ* < *bldK*/*bldL* < *bldA*/*bldH* < *bldG* < *bldC* < *bldD*/*bldM* (Nodwell *et al.*, 1999). Each mutant strain can complement all those listed to its left and can be complemented by those listed to its right. The nature of the diffusing signals has not yet been characterised. Mutations leading to the bald phenotype affect a number of genes involved in very different functions: for example *bldK* encodes an ABC transporter, *bldA* encodes a tRNA and *bldD* encodes a DNA-binding protein.

White Mutants

In *Streptomyces coelicolor*, the spores do not mature completely in "white" mutants (Chater, 1993; Flardh *et al.*, 1999). The *whiG* and *sigF* loci encode sigma factors that control the late stages of sporulation. The expression of *sigF* depends indirectly on the production of WhiG.

Thus, cellular differentiation of *Streptomyces* involves the sequential activation of various sigma factors, each of which directs the transcription of a specific batch of genes. This is also the case in bacteria belonging to the genera *Bacillus* and *Myxococcus*.

Translational Regulation

Regulation at the translational level is also involved in differentiation.The *bldA* locus of "*bald*" mutants (Chater, 1993; Kelemen and Buttner, 1998) specifies a tRNA that recognises the rare leucine codon, UUA (Chater, 1998). This codon (as TTA) is present in many genes that regulate the antibiotic biosynthesis pathways, including *redZ,* which positively controls RedD, the activator of the undecylprodigiosin biosynthetic pathway (White and Bibb, 1997) and *act*II-ORF4, the activator of the actinorhodin biosynthetic pathway (Fernandez-Moreno *et al.*, 1991). Only small amounts of these two antibiotics are found in *bldA* mutants. However, if the TTA codon is replaced by a TTG codon in actII-ORF4 a larger amount of actinorhodin is produced constitutively. Thus, the expression of these genes is regulated by controlling their translation in function of the availability of the tRNA that recognises this rare codon.

Stress and Development

It has long been thought that there is a connection between stress and development in the Streptomycetes. This is has now been confirmed in *S. coelicolor*. Firstly, σ^H was identified as one of the sigma factors associated with the stress response; then it was shown that σ^H is temporally and spatially regulated during the cell cycle. In particular, the tissue specific expression of σ^H was shown to be mediated by the transcription factor, BldD. *bldD* was one of the first "bald" developmental genes isolated from *Streptomyces*. These results indicate that stress and development programmes are closely linked (Kelemen *et al.*, 2001).

Regulation of Morphological Differentiation

Nutritional deficiencies probably initiate the differentiation process.

Some of the signals that trigger differentiation have been identified. For example, an amino acid deficiency in nitrogen-limiting conditions may induce the response and trigger the synthesis of ppGpp. Morphological differentiation is less marked and the amount of antibiotics synthesised is much lower in a strain harbouring a mutation in the *relA* gene (encoding a ppGpp synthase). The synthesis of ppGpp is thus required for morphological and metabolic differentiation in *Streptomyces* (Chakraburtty and Bibb, 1997; Takano *et al.*, 2001). It is also known that phosphate deficiency strongly stimulates the production of antibiotics and sporulation (Doull and Vining, 1990). In addition, small diffusible molecules, γ-butyrolactones, are produced by many species of *Streptomyces* during late stages of development. These molecules are involved in the development of the aerial mycelium and the production of antibiotics. They may also be involved in communication between cells, thereby facilitating the synchronisation of development (Ohnishi *et al.*, 1999). It has also been shown that the development of the aerial mycelium requires the partial degradation of the basal mycelium, in a process likened to programmed cell death in eukaryotes (Miguelez *et al.*, 1999). To adapt to the change in the nature of the nutrients used for growth, the generation of the aerial mycelium must thus be accompanied by a total rearrangement of cell metabolism, which in turns leads to the biosynthesis of antibiotics.

Metabolic Differentiation

Common intermediates from primary metabolism are condensed into more complex structures by defined biochemical pathways during secondary metabolism. The products of secondary metabolism do not appear to have any further function in metabolism even though many of them are bioactive molecules. Due to the industrial importance of *Streptomyces* the genetics of secondary metabolites biosynthesis has received a lot of attention (Baltz, 1998; Hopwood *et al.*, 1995).

Genes Involved in Secondary Metabolism

The *Streptomyces* genes involved in the biosynthesis of a given secondary metabolism are clustered. Genes conferring resistance and pathway-specific regulators are often found in these clusters. This clustering facilitated the isolation of genes involved in secondary metabolism and their subsequent genetic study: when a gene involved in resistance or biosynthesis was isolated, the other genes from the same pathway could be found in the vicinity of the first one.

S. coelicolor was known to produce three antibiotics, the production pathways of which are chromosomally encoded: the red pigmented tripyrrole undecylprodigiosin (Red) (Cerdeno *et al.*, 2001), the blue pigmented polyketide actinorhodin (Act) (Fernandez-Moreno *et al.*, 1994; Malpartida and Hopwood, 1986) and the lipopeptide calcium-dependent antibiotic (CDA) (Chong *et al.*, 1998; Hopwood and Wright, 1983). However, the genome sequence revealed over 20 gene clusters potentially able to synthesise very diverse secondary metabolites. In some cases, the gene sequence enabled the type of molecule produced to be predicted (Challis and Ravel, 2000). A similar situation was encountered for the avermectin-producing strain of *S. avermitilis* where sequence analysis has identified 25 gene clusters encoding secondary metabolites. Some of these gene clusters, in both streptomycetes, are thought to be involved in the synthesis of pigments, siderophores, polyketide and peptide compounds (Omura *et al.*, 2001). Secondary metabolite clusters account for about 5 % of the total genome length in *S. coelicolor* and more than 6% in *S. avermitilis*. Therefore, the metabolic capacities of *Streptomyces* are probably more diverse than suggested by biochemical screening programmes. The genetic screening of previously isolated strains could lead to the isolation of new clusters, which could be expressed in controlled conditions to yield new molecules.

In *S. avermitilis* and *S. coelicolor*, many of the secondary metabolism gene clusters are found in the terminal and intermediate part of the chromosome rather than in the centre (Bentley, 2001; Omura *et al.*, 2001).

Regulation of Secondary Metabolism

The expression of genes involved in secondary metabolism is generally tightly regulated and is often associated with the later stages of development (aerial mycelium formation and sporulation on solid media, late exponential phase and stationary phase in liquid culture). Several different classes of genes involved in the regulation of this expression have been identified. Pathway-specific regulators, encoded by genes located in the cluster of biosynthetic genes, control the expression of a given pathway. Pleiotropic regulators can control the production of several secondary metabolites, and some of them can also affect morphological differentiation in addition to secondary metabolism (Bibb, 1996; Hopwood *et al.*, 1995). Much effort is currently being made to elucidate the interactions between these different regulators and their interplay with global regulatory networks. Global transcriptome analysis will be helpful for this type of study, as shown by the results obtained by Huang *et al* (Huang *et al.*, 2001). These authors used DNA microarrays to investigate the transcription of the Red, Act and CDA biosynthetic genes in *S. coelicolor*, and confirmed that antibiotic pathway genes are co-ordinately regulated at the level of transcription. The co-ordinated transcription of contiguous genes allowed the physical boundaries between biosynthetic loci to be predicted more accurately than hitherto and revealed new physical clusters of co-ordinately regulated genes. Additional studies on pathway specific regulators mutants identified chromosomally dispersed sets of genes, the expression of which appears to be regulated together with that of the biosynthetic genes. It should now be possible to investigate the mechanisms responsible for the co-ordinated expression of these genes (pleiotropic regulators, regulation by genes in the biosynthetic pathway, indirect effects of antibiotic production).

Relationship Between Primary and Secondary Metabolism.

The building blocks used to synthesise secondary metabolites are the products of primary metabolism (for a review on *Streptomyces* primary metabolism, see (Hodgson, 2000)). Industrial strains that overproduce antibiotics or other secondary metabolites have been obtained by successive rounds of random mutagenesis. The mutations that increase production may directly affect the genes involved in the biosynthetic pathway of the product in question, as well as the genes from the primary metabolism or its regulation, thus modifying the supply of precursors for secondary metabolism.

DNA microarrays could be used to identify differences in the genomic DNA and in gene expression patterns between overproducing strains and the original wild-type strain. If the purpose is to identify differences affecting housekeeping genes, it might not be necessary to use DNA microarrays corresponding to the strain studied. This was shown by Lum *et al.* who used *S. coelicolor* DNA microarrays to analyse changes in *Saccharopolyspora erythraea* strains (Lum *et al.*, 2001). These authors first checked that the sequences of the housekeeping genes were sufficiently well conserved to allow the use of heterologous DNA arrays. They then compared the transcription patterns of the wild-type *S. erythraea* with that of an erythromycin-overproducing strain to identify genes with modified expression in the industrial strain. The way in which these genes are involved in overproduction needs to be studied. Knowledge concerning the basis of overproduction in existing industrial strains could be used to produce new strains of industrial value. A strain that has been modified to synthesise more of a certain family of secondary metabolites could also serve as a host for the expression of the heterologous biosynthetic clusters that are either newly isolated or come from strains that are less easy to modify by genetic techniques.

Evolution of Secondary Metabolism.

Many conflicting hypotheses, that will not be discussed here, have been formulated concerning the role of secondary metabolism. It has been proposed that secondary metabolism plays a major role in biochemical evolution and that "the easy conditions in the area of secondary metabolism in the biochemical evolution game correspond to some facilities in the genetic field" (Zähner, 1979). When this sentence was written, it was considered that a large part of the genetic information required for secondary metabolism might be carried by plasmids in *Streptomyces* (Hopwood, 1978). This hypothesis was based upon the observation that the methylenomycin biosynthetic genes in *S. coelicolor* were plasmid-borne (Kirby and Hopwood, 1977). It was proposed to explain the genetic instability of antibiotic production in many strains and the fact that the ability to produce a certain secondary metabolite is often strain specific in a given species. We now know that methylenomycin is the exception rather than the rule. However, Zähner's hypothesis still holds true. Genome sequencing has shown that the number of genes potentially involved in secondary metabolism is higher than expected and a lot of these genes are located in the terminal parts of the chromosome. These are regions of enhanced genome plasticity that might undergo frequent genetic exchange and rearrangements, allowing the rapid evolution of the genes harboured. So far the mechanisms leading to the evolution of secondary metabolism genes within or between clusters are unknown. The known clusters are not part of "secondary metabolism islands" with structure similar to pathogenicity islands. However it should be mentioned that in *Streptomyces argillaceus*, the mithramycin cluster is flanked by two direct repeated DNA sequences of 241-bp, suggesting incorporation into the chromosome of the cluster, from a hypothetical circular extrachromosomal element through Campbell recombination (Lombo *et al.*, 1999).

Functional Analysis.

A vast functional analysis programme, entitled the *Streptomyces coelicolor* Investigating Gene Function Initiative (http://gepasi.dbs.aber.ac.uk/S_coeli/home.html or http://qbab.aber.ac.uk/S_coeli/home.html), has been funded by the BBSRC in the UK. Several British laboratories are involved in this programme, the aims of which are to generate *S. coelicolor* mutants and to study the transcriptome and the proteome. Other initiatives have also been undertaken to exploit the *S. coelicolor* sequence data.

Libraries of S. coelicolor Mutants

Even though *S. coelicolor* has been widely studied, only about 200 genes have been inactivated by gene disruption in this strain. Therefore, new genetic tools have been developed to generate libraries of mutants of *S. coelicolor*. The approaches used involve random and targeted mutagenesis.

Creation of Libraries of Mutants by Random Mutagenesis

A *S. coelicolor* library of about 100,000 independent mutants was generated by Tn*4560* mutagenesis (Fowler and Kieser, 2001). The insertion sites are randomly distributed throughout the entire chromosome and there does not appear to be any target sequence specificity. PCR-based techniques can be used to detect the presence of transposons in specific genes and to determine the precise insertion site. Another strategy involves *in vitro* transposon mutagenesis of cosmids from the ordered cosmid library by use of a mini-transposon containing a promoterless copy of *egfp*, the modified gene encoding the green fluorescent protein (Herron *et al.*, 2001). Transposon-containing cosmids are introduced into *S. coelicolor* and mutants are obtained by gene replacement. The transcriptional or translational fusion of *egfp* allows the expression of the disrupted gene to be analysed and can be used to determine the cellular location of the fusion protein obtained.

Creation of Libraries of Mutants by Targeted Mutagenesis

It has been possible to carry out directed mutagenesis in *Streptomyces* for 20 years but this is a very time consuming process. This is demonstrated by the relatively small number of mutants described even in *S. coelicolor*. Long periods of time were required to construct the plasmid and to search for the double crossover events leading to a stable "knock-out" mutant. However, the availability of the whole *S. coelicolor* sequence and of systems facilitating targeted gene disruption in *E. coli* have dramatically improved the targeted mutagenesis process in *Streptomyces*. It is now possible to use homologous recombination to specifically disrupt any gene carried on a cosmid. This involves simple transformation with linear DNA (PCR) in an *ad hoc E. coli* background, such as a strain overexpressing the *exo, bet* and *gam* genes of phage lambda (Chaveroche *et al.*, 2000; Datsenko and Wanner, 2000). Such cosmids, containing large regions of homologous DNA on both sides of the mutated gene, can then be introducted into *Streptomyces* by transformation or conjugation. As these cosmids are unable to replicate in *Streptomyces,* integration is selected, but due to the large homologous regions on both sides of the mutated gene, double crossover events (marker exchange) are very likely.

This technique should facilitate genetic studies on *S. coelicolor* but could also be applied in any microorganism where a cosmid library and the genome sequence are available.

PhiC31-derived phage vectors deleted for the phage attachment site have been used for gene disruption experiments at the John Innes Centre (Chater and Bruton, 1985; Malpartida and Hopwood, 1986; Smith *et al.*, 1999). The great interest of this method is that DNA is delivered through phage infection, a natural process avoiding protoplasting, which is known to be mutagenic.

The two main drawbacks are that the strain must be susceptible to this phage and that the size of the phage-packaging is limited. The maximum size of the insert is 6 kb, which is adequate for single crossover gene disruption, but a real constraint for double crossover gene replacements.

Transcriptome Studies

As soon as a major part of the *S. coelicolor* sequence became available, different groups began to study the transcriptome. C. Smith's group in Manchester (http://www.arrays.umist.ac.uk) co-ordinated a UK programme to study the functional genomics of *S. coelicolor* and the groups of S. N. Cohen and C. Kao in Stanford also developped DNA arrays that they used for the study of *S. coelicolor* (Huang *et al.*, 2001) and other species (Lum *et al.*, 2001). In both cases, PCR primers were designed from the emerging genome. A first version of microarrays, containing about 75 % of the genes *of S. coelicolor,* was obtained by these groups and whole genome arrays should be obtained rapidly. The microarrays from the Manchester group should be available from MWG biotech UK.

Proteome Studies

In the early 1990s, Charles Thompson and co-workers used two-dimensional gel electrophoresis to study protein production in *S. coelicolor* (Puglia *et al.*, 1995). They showed that the growth curve of *S. coelicolor* could be separated into four components: rapid growth 1, transition, rapid growth 2 and stationary phase. Proteins extracted from radiolabelled liquid cultures at 16 intervals during development were separated and quantified on 2-D gel. Statistical comparisons of 2-D gels patterns found four distinct 2-D patterns, each specific for a different developmental stage. A recent study of prokaryotic development used 2D gels analyses to identify stress-induced regulatory networks (Vohradsky *et al.*, 2000). All available 2-D gel data are presented on the SwiCz web site maintained in Prague (http://proteom.biomed.cas.cz/). A total of 1237 protein spots have been identified, and the developmental expression profile of each is available. The protein spots have been grouped into clusters presenting similar patterns. By November 2001 123 proteins had been identified, but this number should rapidly increase.

The *S. coelicolor* proteome is also being analysed at the John Innes Centre. MALDI-TOF peptide mass fingerprinting has been used on 550 spots visualised on 2D gels and 481 could be identified. Changes in protein production at different stages of growth and in mutants (e.g. sigma factor mutant) will also be studied. Analytical tools for this purpose are being developed at the University of Wales at Aberystwyth (http://gepasi.dbs.aber.ac.uk/S_coeli/home.html).

Conclusion

We hope that we have convinced the reader that *Streptomyces* is an advanced bacterium by any standards. The availability of the sequence, genetic and molecular tools, know-how and funding should allow the numerous dedicated teams to obtain a more comprehensive picture of this sophisticated bacterium.

Acknowledgement

We are grateful to David Hopwood for encouragement, critical reading of our manuscript and for communication of unpublished results.

Note Added in Proof

Since this manuscript was submitted, the sequence of *Streptomyces coelicolor* has been published (Bentley, S.D. *et al.*, 2002. Complete genome sequence of the model actinomycete *Streptomyces coelicolor* A3(2). Nature. 417: 141-147).

References

Altenbuchner, J. and Cullum, J. 1984. DNA amplification and an unstable arginine gene in *Streptomyces lividans* 66. Mol. Gen. Genet. 195: 134-138.

Baltz, R.H. 1998. Genetic manipulation of antibiotic-producing *Streptomyces*. Trends Microbiol. 6: 76-83.

Bao, K. and Cohen, S.N. 2001. Terminal proteins essential for the replication of linear plasmids and chromosomes in Streptomyces. Genes Dev. 15: 1518-1527.

Bentley, S. 2001. The *Streptomyces coelicolor* A3(2) genome project at the

Sanger Centre. *In* "12th International Symposium on the Biology of Actinomycetes" , pp. 47, Vancouver.

Bibb, M. 1996. 1995 Colworth Prize Lecture. The regulation of antibiotic production in *Streptomyces coelicolor* A3(2). Microbiology. 142: 1335-1344.

Bibb, M.J., Ward, J.M., Kieser, T., Cohen, S.N. and Hopwood, D.A. 1981. Excision of chromosomal DNA sequences from Streptomyces coelicolor forms a novel family of plasmids detectable in Streptomyces lividans. Mol. Gen. Genet. 184: 230-240.

Boccard, F., Smokvina, T., Pernodet, J.L., Friedmann, A. and Guerineau, M. 1989. The integrated conjugative plasmid pSAM2 of *Streptomyces ambofaciens* is related to temperate bacteriophages. EMBO J. 8: 973-980.

Bourn, W.R. and Babb, B. 1995. Computer assisted identification and classification of Streptomycete promoters. Nucl. Acids Res. 23: 3696-3703.

Cerdeno, A.M., Bibb, M.J. and Challis, G.L. 2001. Analysis of the prodiginine biosynthesis gene cluster of *Streptomyces coelicolor* A3(2): new mechanisms for chain initiation and termination in modular multienzymes. Chem. Biol. 8: 817-829.

Chakraburtty, R. and Bibb, M. 1997. The ppGpp synthetase gene (*relA)* of *Streptomyces coelicolor* A3(2) plays a conditional role in antibiotic production and morphological differentiation. J. Bacteriol. 179: 5854-5861.

Challis, G.L. and Ravel, J. 2000. Coelichelin, a new peptide siderophore encoded by the *Streptomyces coelicolor* genome: structure prediction from the sequence of its non-ribosomal peptide synthetase. FEMS Microbiol. Lett. 187: 111-114.

Chang, P.C. and Cohen, S.N. 1994. Bidirectional replication from an internal origin in a linear *Streptomyces* plasmid. Science. 265: 952-954.

Chang, P.C., Kim, E.S. and Cohen, S.N. 1996. *Streptomyces* linear plasmids that contain a phage-like, centrally located, replication origin. Mol. Microbiol. 22: 789-800.

Chater, K. 1998. Taking a genetic scalpel to the *Streptomyces* colony. Microbiology. 144: 1465-1478.

Chater, K.F. 1993. Genetics of differentiation in *Streptomyces.* Annu. Rev. Microbiol. 47: 685-713.

Chater, K.F. 2001. Regulation of sporulation in *Streptomyces coelicolor* A3(2): a checkpoint multiplex? Curr. Opin. Microbiol. 4: 667-673.

Chater, K.F. and Bruton, C.J. 1985. Resistance, regulatory and production genes for the antibiotic methylenomycin are clustered. EMBO J. 4: 1893-1897.

Chaveroche, M.K., Ghigo, J.M. and d'Enfert, C. 2000. A rapid method for efficient gene replacement in the filamentous fungus *Aspergillus nidulans.*

Nucl. Acids Res. 28: E97.

Chen, C. 1996. Complications and implications of linear bacterial chromosomes. Trends Genet. 12: 192-196.

Chong, P.P., Podmore, S.M., Kieser, H.M., Redenbach, M., Turgay, K., Marahiel, M., Hopwood, D.A. and Smith, C.P. 1998. Physical identification of a chromosomal locus encoding biosynthetic genes for the lipopeptide calcium-dependent antibiotic (CDA) of *Streptomyces coelicolor* A3(2). Microbiology. 144: 193-199.

Cole, S.T., Brosch, R., Parkhill, J., Garnier, T., Churcher, C., Harris, D., Gordon, S.V., Eiglmeier, K., Gas, S., Barry, C.r., Tekaia, F., Badcock, K., Basham, D., Brown, D., Chillingworth, T., Connor, R., Davies, R., Devlin, K., Feltwell, T., Gentles, S., Hamlin, N., Holroyd, S., Hornsby, T., Jagels, K., Barrell, B.G. and et, a.l. 1998. Deciphering the biology of *Mycobacterium tuberculosis* from the complete Nature. 393: 537-544.

Dary, A., Martin, P., Wenner, T., Decaris, B. and Leblond, P. 2000. DNA rearrangements at the extremities of the *Streptomyces ambofaciens* linear chromosome: evidence for developmental control. Biochimie. 82: 29-34.

Datsenko, K.A. and Wanner, B.L. 2000. One-step inactivation of chromosomal genes in *Escherichia coli* K-12 using PCR products. Proc. Natl. Acad. Sci. USA. 97: 6640-6645.

Doull, J.L. and Vining, L.C. 1990. Nutritional control of actinorhodin production by *Streptomyces coelicolor* A3(2): suppressive effects of nitrogen and phosphate. Appl Microbiol. Biotechnol. 32: 449-454.

Fernandez-Moreno, M.A., Caballero, J.L., Hopwood, D.A. and Malpartida, F. 1991. The *act* cluster contains regulatory and antibiotic export genes, direct targets for translational control by the *bldA* tRNA gene of *Streptomyces*. Cell. 66: 769-780.

Fernandez-Moreno, M.A., Martinez, E., Caballero, J.L., Ichinose, K., Hopwood, D.A. and Malpartida, F. 1994. DNA sequence and functions of the *actVI* region of the actinorhodin biosynthetic gene cluster of *Streptomyces coelicolor* A3(2). J. Biol. Chem. 269: 24854-24863.

Fischer, G., Kyriacou, A., Decaris, B. and Leblond, P. 1997. Genetic instability and its possible evolutionary implications on the chromosomal structure of *Streptomyces*. Biochimie. 79: 555-558.

Fischer, G., Wenner, T., Decaris, B. and Leblond, P. 1998. Chromosomal arm replacement generates a high level of intraspecific polymorphism in the terminal inverted repeats of the linear chromosomal DNA of *Streptomyces ambofaciens*. Proc. Natl. Acad. Sci. USA. 95: 14296-14301.

Flardh, K., Findlay, K.C. and Chater, K.F. 1999. Association of early sporulation genes with suggested developmental decision points in *Streptomyces coelicolor* A3(2). Microbiology. 145: 2229-2243.

Flett, F. and Cullum, J. 1987. DNA deletions in spontaneous chloramphenicol-sensitive mutants of *Streptomyces coelicolor* A 3(2) and *Streptomyces lividans* 66. Mol. Gen. Genet. 207: 499-502.

Flett, F., Mersinias, V. and Smith, C.P. 1997. High efficiency intergeneric conjugal transfer of plasmid DNA from *Escherichia coli* to methyl DNA-restricting streptomycetes. FEMS Microbiol. Lett. 155: 223-229.

Fowler, K. and Kieser, T. 2001. The *Streptomyces coelicolor* A3(2) transposon mutant "Megalibrary" : a valuable tool for investigating gene function. *In* "12th International Symposium on the Biology of Actinomycetes". pp. 81. Vancouver.

Gravius, B., Glocker, D., Pigac, J., Pandza, K., Hranueli, D. and Cullum, J. 1994. The 387 kb linear plasmid pPZG101 of *Streptomyces rimosus* and its interactions with the chromosome. Microbiology. 140: 2271-2277.

Hayakawa, T., Otake, N., Yonehara, H., Tanaka, T. and Sakaguchi, K. 1979a. Isolation and characterization of plasmids from *Streptomyces*. J. Antibiot. (Tokyo). 32: 1348-1350.

Hayakawa, T., Tanaka, T., Sakaguchi, K., Otake, N. and Yonehara, H. 1979b. A linear plasmid-like DNA in *Streptomyces* sp. producing lankacidin group antibiotics. J. Gen. Appl. Microbiol. 25: 255-260.

Herron, P., Fielding, S., del Sol, R., Bishop, A., Evans, M. and Dyson, P. 2001. Transposon mutagenesis of *Streptomyces coelicolor* A3(2). *In* "12th International Symposium on the Biology of Actinomycetes". pp. 151. Vancouver.

Hodgson, D.A. 2000. Primary metabolism and its control in Streptomycetes: a most unusual group of bacteria. Adv. Microb. Physiol. 42: 47-238.

Hopwood, D. 1965. A circular linkage map in the actinimycete *Streptomyces coelicolor*. J. Mol. Biol. 12: 514-516.

Hopwood, D.A. 1978. Extrachromosomally determined antibiotic production. Ann. Rev. Microbiol. 32: 373-392.

Hopwood, D.A. 1999. Forty years of genetics with *Streptomyces*: from *in vivo* through *in vitro* to *in silico*. Microbiology. 145: 2183-2202.

Hopwood, D.A., Chater, K.F. and Bibb, M.J. 1995. Genetics of antibiotic production in *Streptomyces coelicolor* A3(2), a model streptomycete. Biotechnology. 28: 65-102.

Hopwood, D.A., Hintermann, G., Kieser, T. and Wright, H.M. 1984. Integrated DNA sequences in three streptomycetes form related autonomous plasmids after transfer to *Streptomyces lividans*. Plasmid. 11: 1-16.

Hopwood, D.A., Kieser, T., Wright, H.M. and Bibb, M.J. 1983. Plasmids, recombination and chromosome mapping in *Streptomyces lividans* 66. J. Gen. Microbiol. 129: 2257-2269.

Hopwood, D.A. and Wright, H.M. 1978. Bacterial protoplast fusion:

recombination in fused protoplasts of *Streptomyces coelicolor*. Mol. Gen. Genet. 162: 307-317.

Hopwood, D.A. and Wright, H.M. 1983. CDA is a new chromosomally-determined antibiotic from *Streptomyces coelicolor* A3(2). J. Gen. Microbiol. 129: 3575-3579.

Hopwood, D.A., Wright, H.M., Bibb, M.J. and Cohen, S.N. 1977. Genetic recombination through protoplast fusion in *Streptomyces*. Nature. 268: 171-174.

Huang, C.-H., Lin, Y.-S., Yang, Y.-L., Huang, S.-W. and Chen, C.W. 1998. The telomeres of *Streptomyces* chromosomes contain conserved palindromic sequences with potential to form complex secondary structures. Mol. Microbiol. 28: 905-926.

Huang, J., Lih, C.J., Pan, K.H. and Cohen, S.N. 2001. Global analysis of growth phase responsive gene expression and regulation of antibiotic biosynthetic pathways in *Streptomyces coelicolor* using DNA microarrays. Genes Dev. 15: 3183-3192.

Kelemen, G.H. and Buttner, M.J. 1998. Initiation of aerial mycelium formation in *Streptomyces*. Curr. Opin. Microbiol. 1: 656-662.

Kelemen, G.H., Viollier, P.H., Tennor, J., Marri, L., Buttner, M.J. and Thompson, C.J. 2001. A connection between stress and development in the multicellular prokaryote *Streptomyces coelicolor* A3(2). Mol. Microbiol. 40: 804-814.

Kieser, H.M., Kieser, T. and Hopwood, D.A. 1992. A combined genetic and physical map of the *Streptomyces coelicolor* A3(2) chromosome. J. Bacteriol. 174: 5496-5507.

Kinashi, H., Shimaji, M. and Sakai, A. 1987. Giant linear plasmids in *Streptomyces* which code for antibiotic biosynthesis genes. Nature. 328: 454-456.

Kinashi, H., Shimaji-Murayama, M. and Hanafusa, T. 1992. Integration of SCP1, a giant linear plasmid, into the *Streptomyces coelicolor* chromosome. Gene. 115: 35-41.

Kirby, R. and Hopwood, D.A. 1977. Genetic determination of methylenomycin synthesis by the SCP1 plasmid of *Streptomyces coelicolor* A3(2). J. Gen. Microbiol. 98: 239-252.

Leblond, P., Demuyter, P., Simonet, J.M. and Decaris, B. 1990. Genetic instability and hypervariability in *Streptomyces ambofaciens*: towards an understanding of a mechanism of genome plasticity. Mol. Microbiol. 4: 707-714.

Leblond, P., Fischer, G., Francou, F.X., Berger, F., Guerineau, M. and Decaris, B. 1996. The unstable region of *Streptomyces ambofaciens* includes 210 kb terminal inverted repeats flanking the extremities of the linear chromosomal DNA. Mol. Microbiol. 19: 261-271.

Lezhava, A., Mizukami, T., Kajitani, T., Kameoka, D., Redenbach, M., Shinkawa, H., Nimi, O. and Kinashi, H. 1995. Physical map of the linear chromosome of *Streptomyces griseus*. J. Bacteriol. 177: 6492-6498.

Lin, Y.-S., Kieser, H.M., Hopwood, D.A. and Chen, C.W. 1993. The chromosomal DNA of *Streptomyces lividans* 66 is linear. Mol. Microbiol. 10: 923-933.

Lombo, F., Brana, A.F., Mendez, C. and Salas, J.A. 1999. The mithramycin gene cluster of *Streptomyces argillaceus* contains a positive regulatory gene and two repeated DNA sequences that are located at both ends of the cluster. J. Bacteriol. 181: 642-647.

Lonetto, M.A., Brown, K.L., Rudd, K.E. and Buttner, M.J. 1994. Analysis of the *Streptomyces coelicolor sigE* gene reveals the existence of a subfamily of eubacterial RNA polymerase σ factors involved in the regulation of extracytoplasmic functions. Proc. Natl. Acad. Sci. USA. 91: 7573-7577.

Lum, A.M., Kao, C.M., Huang, J. and Cohen, S.N. 2001. DNA microarray analyses of a *Saccharopolyspora erythraea* overproducer strain *In* "12th International Symposium on the Biology of Actinomycetes". pp. 61. Vancouver.

MacNeil, D.J., Gewain, K.M., Ruby, C.L., Dezeny, G., Gibbons, P.H. and MacNeil, T. 1992. Analysis of *Streptomyces avermitilis* genes required for avermectin biosynthesis utilizing a novel integration vector. Gene. 111: 61-68.

Malpartida, F. and Hopwood, D.A. 1986. Physical and genetic characterisation of the gene cluster for the antibiotic actinorhodin in *Streptomyces coelicolor* A3(2). Mol. Gen. Genet. 205: 66-73.

Miguelez, E.M., Hardisson, C. and Manzanal, M.B. 1999. Hyphal death during colony development in *Streptomyces antibioticus*: morphological evidence for the existence of a process of cell deletion in a multicellular prokaryote. J. Cell. Biol. 145: 515-525.

Musialowski, M.S., Flett, F., Scott, G.B., Hobbs, G., Smith, C.P. and Oliver, S.G. 1994. Functional evidence that the principal DNA replication origin of the *Streptomyces coelicolor* chromosome is close to the *dnaA-gyrB* region. J. Bacteriol. 176: 5123-5125.

Nodwell, J.R., Yang, M., Kuo, D. and Losick, R. 1999. Extracellular complementation and the identification of additional genes involved in aerial mycelium formation in *Streptomyces coelicolor*. Genetics. 151: 569-584.

Ohnishi, Y., Kameyama, S., Onaka, H. and Horinouchi, S. 1999. The A-factor regulatory cascade leading to streptomycin biosynthesis in *Streptomyces griseus* : identification of a target gene of the A-factor receptor. Mol. Microbiol. 34: 102-111.

Omura, S., Ikeda, H., Ishikawa, J., Hanamoto, A., Takahashi, C., Shinose, M., Takahashi, Y., Horikawa, H., Nakazawa, H., Osonoe, T., Kikuchi, H., Shiba, T., Sakaki, Y. and Hattori, M. 2001. Genome sequence of an industrial microorganism *Streptomyces avermitilis*: deducing the ability of producing secondary metabolites. Proc. Natl. Acad. Sci. USA. 98: 12215-12220.

Puglia, A.-M., Vohradsky, J. and Thompson, C.J. 1995. Developmental control of the heat-shock stress regulon in *Streptomyces coelicolor*. Mol. Microbiol. 17: 737-746.

Qin, Z. and Cohen, S. 1998. Replication at the telomeres of the linear plasmid pSLA2. Mol. Microbiol. 28: 893-904.

Rauland, U., Glocker, I., Redenbach, M. and Cullum, J. 1995. DNA amplifications and deletions in *Streptomyces lividans* 66 and the loss of one end of the linear chromosome. Mol. Gen. Genet. 246: 37-44.

Redenbach, M., Bibb, M.J., Gust, B., Seitz, B. and Spychaj, A. 1999. The linear plasmid SCP1 of *Streptomyces coelicolor* A3(2) possesses a centrally located origin and shows significant homology to the transposon Tn*4811*. Plasmid. 42: 174-185.

Redenbach, M., Flett, F., Piendl, W., Glocker, I., Rauland, U., Wafzig, O., Kliem, R., Leblond, P. and Cullum, J. 1993. The *Streptomyces lividans* 66 chromosome contains a 1 MB deletogenic region flanked by two amplifiable regions. Mol. Gen. Genet. 241: 255-262.

Redenbach, M., Kieser, H.M., Denapaite, D., Eichner, A., Cullum, J., Kinashi, H. and Hopwood, D.A. 1996. A set of ordered cosmids and a detailed genetic and physical map for the 8 Mb *Streptomyces coelicolor* A3(2) chromosome. Mol. Microbiol. 21: 77-96.

Redenbach, M., Scheel, J. and Schmidt, U. 2000. Chromosome topology and genome size of selected actinomycetes species. Antonie Van Leeuwenhoek. 78: 227-235.

Reeves, A.R., Post, D.A. and Vanden Boom, T.J. 1998. Physical-genetic map of the erythromycin-producing organism *Saccharopolyspora erythraea*. Microbiology. 144: 2151-2159.

Schneider, D., Bruton, C.J. and Chater, K.F. 2000. Duplicated gene clusters suggest an interplay of glycogen and trehalose metabolism during sequential stages of aerial mycelium development in *Streptomyces coelicolor* A3(2). Mol. Gen. Genet. 263: 543-553.

Simonet, J.-M., Schneider, D., Volff, J.-N., Dary, A. and Decaris, B. 1992. Genetic instability in *Streptomyces ambofaciens*: inducibility and associated genome plasticity. Gene. 115: 49-54.

Smith, M.C., Burns, R.N., Wilson, S.E. and Gregory, M.A. 1999. The complete genome sequence of the *Streptomyces* temperate phage straight phiC31: evolutionary relationships to other viruses. Nucl. Acids Res. 27:

2145-2155.
Stahl, F. 1967. Circular genetic maps. J. Cell Physiol. 70: 1-12.
Stahl, F. and Steinberg, C. 1964. The theory of formal phage genetics for circular maps. Genetics. 50: 531-538.
Strohl, W.R. 1992. Compilation and analysis of DNA sequences associated with apparent streptomycete promoters. Nucl. Acids Res. 20: 961-974.
Takano, E., Chakraburtty, R., Nihira, T., Yamada, Y. and Bibb, M.J. 2001. A complex role for the gamma-butyrolactone SCB1 in regulating antibiotic production in *Streptomyces coelicolor* A3(2). Mol. Microbiol. 41: 1015-1028.
Vohradsky, J., Li, X.M., Dale, G., Folcher, M., Nguyen, L., Viollier, P.H. and Thompson, C.J. 2000. Developmental control of stress stimulons in *Streptomyces coelicolor* revealed by statistical analyses of global gene expression patterns. J. Bacteriol. 182: 4979-4986.
Volff, J.N. and Altenbuchner, J. 1998. Genetic instability of the *Streptomyces* chromosome. Mol. Microbiol. 27: 239-246.
Volff, J.N. and Altenbuchner, J. 2000. A new beginning with new ends: linearisation of circular chromosomes during bacterial evolution. FEMS Microbiol. Lett. 186: 143-150.
Volff, J.N., Viell, P. and Altenbuchner, J. 1997. Artificial circularization of the chromosome with concomitant deletion of its terminal inverted repeats enhances genetic instability and genome rearrangement in *Streptomyces lividans*. Mol. Gen. Genet. 253: 753-760.
Waldron, C., Matsushima, P., Rosteck, P.R., Jr., Broughton, M.C., Turner, J., Madduri, K., Crawford, K.P., Merlo, D.J. and Baltz, R.H. 2001. Cloning and analysis of the spinosad biosynthetic gene cluster of *Saccharopolyspora spinosa*. Chem. Biol. 8: 487-499.
Wang, S.J., Chang, H.M., Lin, Y.S., Huang, C.H. and Chen, C.W. 1999. *Streptomyces* genomes: circular genetic maps from the linear chromosomes. Microbiology. 145: 2209-2220.
White, J. and Bibb, M. 1997. *bldA* dependence of undecylprodigiosin production in *Streptomyces coelicolor* A3(2) involves a pathway-specific regulatory cascade. J. Bacteriol. 179: 627-633.
Willey, J., Schwedock, J. and Losick, R. 1993. Multiple extracellular signals govern the production of a morphogenetic protein involved in aerial mycelium formation by *Streptomyces coelicolor*. Genes Dev. 7: 895-903.
Yang, M.C. and Losick, R. 2001. Cytological evidence for association of the ends of the linear chromosome in *Streptomyces coelicolor*. J. Bacteriol. 183: 5180-5186.
Zähner, H. 1979. What are secondary metabolites? Folia Microbiol. 24: 435-443.

From: *Genomics of GC-Rich Gram-Positive Bacteria*

5

Impact of the *Mycobacterium leprae* Genome Sequence on Leprosy Research

Varalakshmi D. Vissa and Patrick J. Brennan

Abstract

The genome of *Mycobacterium leprae*, the causative agent of leprosy, is a product of extensive reductive evolution, with the capabilities of expressing a mere ~1600 proteins. It is anticipated that comparison of those proteins with those encoded by *Mycobacterium tuberculosis*, *Mycobacterium avium* and others, will lead to a set of proteins/peptides suitable for the specific diagnosis of leprosy. Analysis of the transcriptome through microarray technology should indicate whether any of the large array of pseudogenes are transcribed. The genome has demonstrated defects in catabolic pathways, their regulation and in transport systems and thus provides a metabolic rationale for the obligate intracellularism of *M. leprae* and perhaps a blueprint for rectifying these defects leading to ultimate growth *in vitro*. New evidence for allelic diversity in the genome holds promise for the epidemiological tracking of individual strains. The messages being derived from a genome unique in its lack of gene redundancy, dominance of pseudogenes and massive defects in critical pathways are profound and could galvanize leprosy research towards answering age-old questions, before the disease itself fades from public consciousness.

Introduction

Leprosy, an infectious disease primarily of the skin and nerves, is caused by *Mycobacterium leprae* whose genomic sequence has now been revealed (Cole *et al.*, 2001). In this chapter, we discuss some of the future directions in which this new knowledge may be applied in efforts to eliminate leprosy, especially in light of studies with other completed genomes and newer technologies. In particular, we compare the information with that being generated in the study of tuberculosis following the availability of the genomic sequence of *Mycobacterium tuberculosis* (Cole *et al.*, 1998). The genus *Mycobacterium* also includes other human pathogens such as *Mycobacterium avium, Mycobacterium marinum*, *Mycobacterium ulcerans*, and the avirulent environmental organism *Mycobacterium smegmatis* (Smida *et al.*, 1988), and members of the *Corynebacterium, Nocardia*, and *Rhodocococcus* genera are related organisms belonging to the common lineage of *Actinomycetales* (order) and *Corynebacterineae* (suborder). *M. leprae* is a slow growing obligate intracellular pathogen. The average doubling time is 12-14 days, while it is ~24 hours for *M. tuberculosis* (Levy, 1976). The reservoir of *M. leprae* is not known, but it has been detected in naturally infected armadillos and sooty mangabey monkeys. In experimental models, *M. leprae* can infect mice and guinea pigs (Krahenbuhl and Adams, 2000). The route of transmission of leprosy is not clear, but is suspected to be via the respiratory route, though infection via broken skin remains a possibility. The primary tissues infected by *M. leprae* are usually the superficial sites of skin and nerve due to its preference for a lower temperature; however, other internal organs can also be infected in uncontrolled disease (Shepard, 1965). The severity and pathology of human leprosy depend on the immunity of the infected persons and have been described by Ridley and Jopling as a continuous spectrum of disease states (Ridley and Jopling, 1966). At one pole of the spectrum, patients with good T-cell mediated (Th1 type) immunity exhibit "tuberculoid" leprosy (TT), while, at the other pole, those with poor T-cell mediated immunity manifest "lepromatous" leprosy (LL). Borderline tuberculoid (BT), borderline-borderline (BB), and borderline lepromatous (BL) states are in between. Changes in immunity and therapy can shift the disease up or down the spectrum. Nerve damage occurs in most cases of *M. leprae* infection at all levels of the disease spectrum (Ganapati and Revankar, 1989). In tuberculoid leprosy, the sensory and motor nerves within the skin are affected, resulting in loss of sensation to stimuli such as touch and heat, while damage of the deeper cutaneous nerve trunks leads to loss of muscle function (paralysis) in LL. Loss of fingers and toes, the characteristic disfiguration in leprosy patients, is due to osteoporosis resulting from inflammation, trauma, and secondary bacterial infections. The nerves of the TT skin lesions show loss of structural integrity, presence of macrophage granulomas with few acid-fast bacilli (AFB) in the Schwann cells, macrophages, or nerves. In LL, there is less involvement of the nerves of the skin; nerves trunks look intact, with little evidence of granuloma formation. High titers of *M. leprae* AFB are found in the Schwann cells, perineurium, nerves, and macrophages. Thus nerve damage is more common in TT and BT patients than LL patients. However, when a sudden onset of the immune reactional states occurs in TT/BT or LL patients, painful swelling(s) of the affected nerve occurs. These are called type I reversal reaction (RR) and type II erythema nodosum leprosum (ENL). The triggers for such an acute phase are not known. In the worst untreated cases, the inflammation causes caseous necrosis of the nerve tissue and irreversible paralysis.

Leprosy is a treatable disease when identified early. There are several criteria including physical evaluation of the number of lesions and involvement of nerves (thickening or loss of sensation to stimuli), bacteriological (enumeration of AFB in skin smears), histopathological (e.g., detection of nerve damage, granuloma) and serological (for antibodies against *M. leprae* antigens) tests. However, facilities and the trained personnel required for the laboratory-based tests are not available at endemic sites in the world. Thus, according to World Health Organization (WHO) recommendations, a simple clinical examination of the number of lesions can be used as a classification tool: <6 as paucibacillary leprosy (PB; includes TT and BT) and ≥ 6 as multibacillary (MB; includes LL, BL and BB patients). Multi drug therapy (MDT), originally designed to prevent the drug resistance evident with dapsone monotherapy, is initiated immediately. The treatment for multibacillary leprosy patients is comprised of monthly, supervised administration of rifampicin (600 mg) and clofazimine (300 mg) and daily doses of self-administered dapsone (100 mg) and clofazimine (50 mg) for a period of one year. For paucibacillary leprosy the regimen consists of monthly doses of rifampicin and daily doses of dapsone for a period of 6 months (World Health Organization Study Group, 1982). The target enzymes for rifampicin and dapsone are the β- subunit of the RNA polymerase and dihydropteroate synthase (*folP1*), respectively. Acquired resistance has been correlated with mutations in the genes of these target enzymes in several bacteria including *M. lepr*ae (Honore and Cole, 1993; Kai *et al.*, 1999, Williams *et al.*, 2000). There is also a report of multidrug resistance in a leprosy patient (Matsuoka *et al.*, 2000a). The target for clofazimine in *M. leprae* has not been determined, but it may bind to DNA and inhibit RNA polymerase (Winder, 1982).

MDT has been very successful in reducing the prevalence rate of global leprosy from 12 cases per 10,000 in 1985 to slightly more than 1 case per 10,000 at the end of 2000. The number of cases registered for treatment globally has declined from 5,351,408 in 1985, to 597,035 as of January 2001. Curiously, the incidence rate (number of new cases detected) had risen from 529,376 in 1995 to a peak of 804,357 in 1998 and subsequently dropped to 719,219 in 2000 (World Health Organization, 2002). There may be several factors contributing to this phenomenon, including the long incubation period of the disease, asymptomatic patient transmitters, leprosy elimination campaigns that are actively detecting cases, and, perhaps, environmental sources of *M. leprae*. Clearly, to win the battle against leprosy, the major research issues that need to be addressed are: epidemiological tracking of clinical cases, early detection, and prevention of drug resistance. In parallel, vaccines and new drugs/regimens may have to be developed. To aid in these studies, it would certainly be beneficial to have methods for the isolation, preservation, propagation, and genetic manipulation of *M. leprae* in a laboratory. It is quite remarkable that we know so much about the organism at the level of its structure (Brennan and Nikaido, 1995), biochemistry (Daffe *et al.*, 1993; Marques *et al.*, 1998), physiology (Wheeler, 1984), drug susceptibility, and genome sequence (Cole *et al.*, 2001), in light of the meager quantities of the bacterium available from infected humans, naturally or intentionally infected armadillos, and, more recently, from the footpads of laboratory nude mice. It is anticipated that the harnessing of the genome information should accelerate this knowledge-acquisition process.

The *M. leprae* Genome

One of several mycobacterial genomes sequenced or being sequenced, that of *M. leprae*, is unique in being the smallest (3.3 Mb), having lower G+C content (~58%), a large number of recombination mediated rearrangements resulting in a mosaic organization, reduced coding capacity (only 50% of the DNA encodes for protein at an average of one gene per 2 kb), and an unusually large number of pseudogenes (~1100). In comparison, the genome of *M. tuberculosis* encodes one gene per kb, nearly 91% of the DNA encodes for putative genes and many instances of gene duplication give rise to considerable redundancy (Cole *et al.*, 1998; Cole *et al.*, 2001). The *M. leprae* genome has resulted from errors in replication combined with a reduced capacity for DNA repair. This reductive evolution, more dramatic than in *Rickettsia prowazekii* in which 24% of the genome represents non-coding DNA and pseudogenes (Andersson *et al.*, 1998), is probably the cause or result of the obligate intracellular habitat of *M. leprae*. In the recently deciphered genome of *Yersinia pestis,* the etiological agent of human plague, 149 pseudogenes (4% of the genome) were identified by comparing against the genome of the closely related enteropathogen, *Yersinia pseudotuberculosis* (Parkhill *et al*., 2001). In particular, the genes lost are those important for the enteropathogenicity of *Y. pseudotuberculosis*, such as surface-expressed proteins involved in motility of the microbe, adhesion, and invasion of the host epithelial cells which are not required for *Y. pestis* for its survival in its non-epithelial cell niche. The ability to obtain metabolites from the host may reduce the need for certain biosynthetic pathways and genes involved in these pathways are not under selection pressure and can be mutated. However, this is not the case in *M. leprae* and *Y. pestis*. Relatively intact biosynthetic pathways in both *M. leprae* and *Y. pestis* compensate for a lack of uptake potential. Several mechanisms have been postulated for gene loss processes (Andersson and Andersson, 2001). For instance, due to the obligate intracellular habitat, there may not be a requirement to adapt to changes in the environment as compared to free-living bacteria, and, therefore, the genetics of regulation need not be very complex. Alternatively, a number of slightly deleterious mutations can arise and accumulate in recurrent bottleneck situations (the "Muller's Ratchet" theory). Lastly, smaller genomes may have a growth advantage.

The *M. leprae* Genome: a Resource for the Elimination of Leprosy

Functional Genomics

Since genomes invariably demonstrate that there are many conserved hypothetical proteins and others that have no known homologs, definition and analysis of the proteome and transcriptome are required to verify and extract useful information about the function of putative genes in a genome, and, of those that have annotated functions, generally, very few are characterized. In the case of *M. leprae*, even though only small amounts of cells were available, Marques *et al.* (Marques *et al*., 1998) detected nearly 400 protein spots representing 25% of the proteome, by two-dimensional electrophoresis of proteins from cytosol, cell wall, and membrane fractions. The identification of proteins excised from two-dimensional gels by peptide mass fingerprinting using LC/MS/MS and the sequence tag method (identification of 4 -5 amino acids at the C-terminal end by micro-sequencing) is quite reliable and sensitive and has been used to derive the published proteome maps of *E. coli* (Tonella *et al*., 1998) and yeast (Sanchez *et al*.,

1996). Since the number of coding sequences (CDSs) is limited (~1600), this method may be economical for the complete definition of the proteome of *M. leprae*. Furthermore, in terms of the source of *M. leprae*, more "viable" cells are now available from mouse footpads (MFP) using improved and well-characterized conditions of isolation and storage (Truman and Krahenbuhl, 2001). As pointed out by Cole *et al.* (Cole *et al.*, 2001), mutations in *metC* in the biosynthetic pathway of methionine and *cysT, cysW, cysA* genes for sulfur transport are likely to render *M. leprae* an auxotroph for methionine and sulfur. Therefore, labeling of MFP-derived *M. leprae* for a prolonged duration with [^{35}S]-methionine should allow the labeling of most proteins for improved sensitivity and detection. In *M. bovis* BCG, the HspX, GroEL-1, GroEL-2, InhA, Tuf, and Rv2623 proteins that are differentially expressed during growth in macrophages were detected by such approaches (Monahan *et al.*, 2001). Using a similar approach, six *M. tuberculosis* proteins that were expressed in macrophages but not in broth culture under normal and stress conditions were detected, and the heat shock proteins, Hsp65 and Hsp70, and members of the antigen 85 complex were found to be upregulated in macrophages in this study (Lee and Horwitz, 1995). Ragno *et al.* (Ragno *et al.*, 2001) have also used similar tools to study the proteins of the macrophage that are regulated upon infection with *M. tuberculosis*. IL-1β and manganese superoxide dismutase were upregulated, but Hsp27 was down regulated. In this study, the mycobacterial protein AtpD, that may be involved in control of cytoplasmic pH, was detected in macrophage extracts. Such information is valuable in defining host-bacterial interactions. Given the limitations of genetic manipulation of *M. leprae*, similar reverse genetic studies are anticipated for its study. Comparison of the 2D-SDS PAGE protein expression profile of *M. leprae* obtained from different sources (human nerve or skin lesions representing the various states of the leprosy spectrum, mouse footpad, armadillo, etc.) should be very informative in identifying virulence/ pathogenic factors and also for development of tools for diagnostics and vaccines.

In terms of the transcriptome, the global expression and regulation of genes of several genomes have been examined by probing immobilized arrays of DNA or oligonucleotides that represent portions of each CDS (microarray technology) with labeled RNA isolated from cells grown under different conditions. For instance, in *M. tuberculosis*, genes that may be involved in the early events of latent tuberculosis have been studied in this way (Sherman *et al.*, 2001). Microarray analysis of transcription in cells grown in a low oxygen environment to mimic conditions that occur in latency revealed that ~60 genes (probably involved in growth in aerobic conditions) were repressed, while 47 other genes were induced. The set of induced genes included those for regulation (a sensor kinase and regulator of a two component regulatory system which, in turn, induce *acr*, the gene for the heat shock protein alpha-crystalline); ion transport (the *bfrB* gene for iron acquisition and *ctpF* for cation transport); detoxification (*ahpC*); nitrogen metabolism (*narX, narK2*); glycolysis (*pfkB*); and a phosphodiesterase (*glpQ1*). Several genes with unknown functions were also found. In *M. leprae*, the orthologs of *bfrB* and *pfkB* are pseudogenes, while, in the other cases, related genes are mutated (*glpQ2, narK3,* and *narG,H,I,J* genes), highlighting the lack of redundancy of *M. leprae*. Microarrays representing the 1604 putative CDS and the ~50 representative pseudogenes of *M. leprae* can be readily designed. The pseudogenes may not be translated, but probing the array with RNA purified from human or animal leprosy infections will indicate whether they are transcribed in *M. leprae*.

In another approach, conditionally essential genes in *M. bovis* BCG were identified by a new method called transposon site hybridization (TraSH) (Sassetti *et al.*, 2001). Genomic DNA of a pool of transposon mutants was obtained, and, after restriction digestion with appropriate enzymes, the DNA flanking the insertion sites was amplified with suitably designed primers. The pooled amplicons were transcribed and labeled *in vitro* and hybridized to a DNA microarray comprised of portions of each putative CDS. By comparing the patterns obtained from cells grown in rich and minimal medium, a few auxotrophic mutations were detected for valine, lecuine, and isoleucine synthesis; sulfate metabolism (resulting in methionine and cysteine auxotrophy); glutamine, glutamate, nitrogen metabolism; and thiamine synthesis. There were mutations in 10 other genes with no putative functions. Similar experiments cannot be conducted in *M. leprae*, since, to date, genetic manipulation has not been feasible. However, the results from this study can be correlated to the mutations in the genome of *M.leprae*. *M. leprae* has pseudogenes in pathways for sulfate transport (*cysT, cysW, cysA*), methionine synthesis (*metC*), glutamine synthesis (*glnA3, glnA4*), and nitrogen regulation (*glnB, glnD, glnQ*) that may have a cumulative effect on the lack of growth *in vitro*. Moreover, the TraSH experiment in *M. bovis* revealed that transpositions did not occur in most genes involved in biosynthetic pathways (amino acids, ribosomal proteins and cell wall synthesis), even when grown in rich medium. This aspect is highlighted in the genome of *M. leprae*, where the pseudogenes are largely in degradation rather than biosynthetic metabolic pathways. Another conclusion is that nutrient acquisition from the environment is probably not very effective even for *M. tuberculosis* and *M. bovis* BCG. It supports the hypothesis that fewer genes for transport are found in *M. tuberculosis* compared to free-living bacteria, and the loss of many of them as pseudogenes in *M. leprae* may not be severely deleterious (Sassetti *et al.*, 2001). However, two transport genes are present in *M. leprae* which do not have orthologs in *M. tuberculosis*: an ABC-sugar transporter and a *Nramp1* like gene for divalent cation acquisition (Cole *et al.*, 2001).

Conditions for Axenic Growth of M. leprae

Development of an *in vitro* method of cultivating *M. leprae* remains an important goal for advancement of leprosy research. *M. leprae* have been shown to be metabolically active *in vitro* for 3-6 weeks in axenic medium, as monitored by the evolution of $[^{14}C]CO_2$ or synthesis of $[^{14}C]$phenolic glycolipid I ($[^{14}C]$PGL-I) from exogenously supplied $[^{14}C]$palmitic acid, but cell division as indicated by ATP levels or doubling of DNA has been shown in very few instances (Bapat, 1989; Dhople *et al.*, 1988a; Nakamura and Matsuoka., 2001; Shannon *et al.*, 1992). Early studies using the techniques of metabolic labeling of whole cells and detection of enzymes in cell extracts focused on identification of metabolic defects that could explain the obligate intracellularism of *M. leprae* and the formulation of a suitable growth medium (Wheeler, 1989a; Wheeler and Ratledge, 1988a). Now that the genome sequence is available and annotated (http://www.sanger.ack.uk/Projects/M leprae/), it is remarkable that most of the earlier results on the peculiarities of *M. leprae* metabolism corroborate with the presence or absence of genes for the various pathways. It is unlikely that conditions that mimic the macrophage environment may be mimicked *in vitro*, using clues from the genome analysis. However, a systematic and comparative approach using supplemental and/or drop out media and a variety of culture conditions may yield results. The genome has uncovered defects in *M. leprae* in mechanisms involved in DNA repair mechanisms, metabolic regulation, transport of amino acids, cations and anions, and in efflux pumps, which cannot be easily corrected in *in vitro* culture. In *M. tuberculosis*, many metabolic pathways have several genes responsible for the same function, either to cope with an extra demand of products or for regulation under different growth conditions (Cole *et al.*, 1998). Such redundancy is lacking in *M. leprae*, thereby reducing its metabolic capacity and flexibility. These may cumulatively explain why *M. leprae* grows very slowly even *in vivo* and hardly at all *in vitro*.

Energy Supply

From the literature and the genome, it is unclear what carbon source *M. leprae* would use in an aerobic or anaerobic respiration mode. There is a complete, but non-redundant complement of genes for glycolysis, tricarboxylic acid cycle (TCA), and hexose monophosphate shunt (HMP) pathways (see Figure 1). *M. leprae* is handicapped in respiration pathways. Genes that encode the NADH dehydrogenase (*ndh1*, the enzyme complex in the oxidative phosphorylation pathway for ATP generation using oxygen as the final acceptor) are defective in *M. leprae*. However, the complexes for oxidation of $FADH_2$ are apparently present. As for anaerobic respiration, with electron acceptors other than oxygen, *M. leprae* does not have functional genes for fumarate reductase (*frdA, frdB, frdC, frdD*), nitrate reductase (*narG, narH, narJ, narI*) or formate dehydrogenase (*fdhD, fdhF*). A *narG* mutant of *M. bovis* BCG was shown to be less virulent compared to a wild type strain in immunodeficient mice (Weber *et al.*, 2000). There may be other electron acceptors, such as dihydroxyphenylalanine (DOPA) for anaerobic respiration, as implicated by the detection of a 3, 4 DOPA oxidase enzyme in whole cells and a particulate fraction of broken *M. leprae* cells isolated from human skin lesions (Prabhakaran, 1967). However, a gene for this function has not been annotated. For these reasons, *M. leprae* may be a microaerophile that is capable of oxygen-dependent growth, but cannot grow in the presence of a level of oxygen equivalent to that present in atmospheric air (21% oxygen). Pyruvate generated from glycolysis can be converted to lactate, ethanol, formate, acetyl CoA, acetate or oxaloacetate (Figure 1). However, the genome indicates that, compared to *M. tuberculosis*, there are fewer genes for these steps in *M. leprae* (Cole *et al.*, 2001). In terms of reactions that recycle NADH back to NAD to keep glycolysis going (and for β -oxidation of lipids), it would be of benefit to drive metabolism towards lactate or ethanol. *M. leprae* uses the anaplerotic enzyme phosphoenolpyruvate carboxylase for funneling oxaloacetate back to the TCA cycle, while *M. tuberculosis* uses pyruvate carboxylase (Cole *et al.*, 2001). There is also a functional glyoxylate cycle in *M. leprae* (again non redundant) that could circumvent two NADH producing reactions of TCA, but at the cost of one GTP and one $FADH_2$ producing step. Is it advantageous to induce this pathway while trying to grow *M. leprae*? The glyoxylate cycle usually takes place when excess acetyl-CoA is introduced into the system from the β -oxidation of lipids. However, there is a marked loss of genes for β-oxidation [*fadD* (acyl-CoA synthases), *fadE* (acyl-CoA dehydrogenases), and *echA* enoyl-CoA hydratase/isomerase superfamily] and, in addition, the pathway for activation of acetate to acetyl-CoA via acetyl-phosphate is impaired due to a pseudogene for the enzyme

Figure 1.

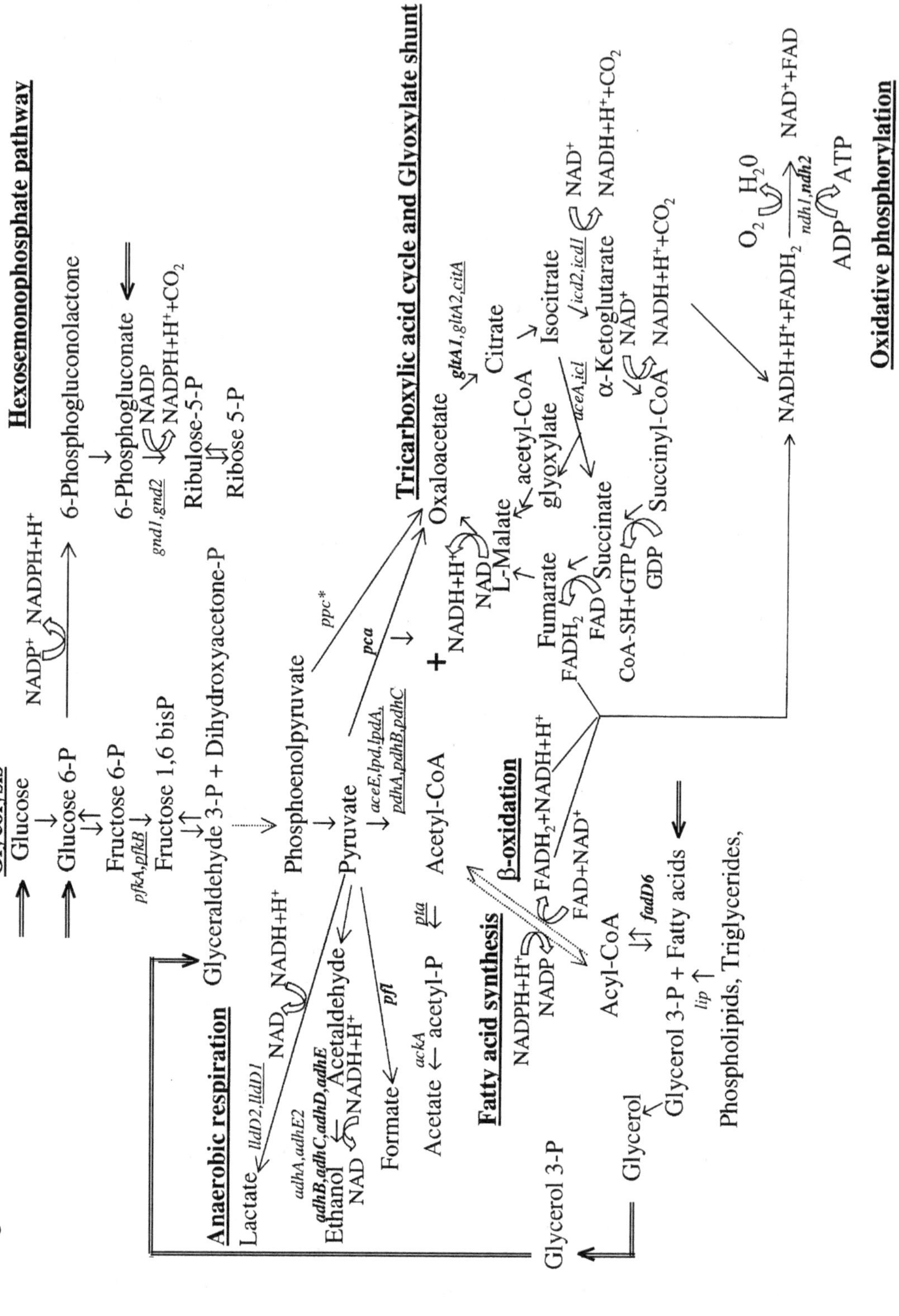
Glycolysis
Glucose
Glucose 6-P
Fructose 6-P
pfkA,pfkB
Fructose 1,6 bisP
Glyceraldehyde 3-P + Dihydroxyacetone-P
Phosphoenolpyruvate
Pyruvate
aceE,lpd,lpdA,
pdhA,pdhB,pdhC
Acetyl-CoA
Hexosemonophosphate pathway
NADP+ NADPH+H+
6-Phosphogluconolactone
6-Phosphogluconate
gnd1,gnd2
NADP
NADPH+H+CO2
Ribulose-5-P
Ribose 5-P
Anaerobic respiration
Lactate
lldD2,lldD1
NAD NADH+H+
adhA,adhE2
adhB,adhC,adhD,adhE
Ethanol
NAD NADH+H+
Acetaldehyde
pfl
Formate
ackA
Acetate ← acetyl-P
pta
ppc*
pca
Tricarboxylic acid cycle and Glyoxylate shunt
Oxaloacetate
gltA1,gltA2,citA
Citrate
Isocitrate
icd2,icd1
NAD+
NADH+H+CO2
α-Ketoglutarate
NAD+
NADH+H+CO2
Succinyl-CoA
CoA-SH+GTP
GDP
Succinate
FAD FADH2
Fumarate
L-Malate
NAD NADH+H+
acetyl-CoA
glyoxylate
aceA,icl
Fatty acid synthesis
NADPH+H+
NADP
β-oxidation
FADH2+NADH+H+
FAD+NAD+
Acyl-CoA
fadD6
Glycerol 3-P + Fatty acids
lip
Phospholipids, Triglycerides,
Glycerol
Glycerol 3-P
Oxidative phosphorylation
O2 H20
NADH+H+FADH2
ndh1,ndh2
NAD+FAD
ADP ATP

phosphotransacetylase (Figure 1) (Cole *et al.*, 2001).

Another issue to be addressed in choosing a carbon source for the possible cultivation of *M. leprae* is the generation of NADPH for fatty acid synthesis and other sugar intermediates such as ribose-5-phosphate for nucleic acid synthesis and arabinose-5-phosphate for synthesis of the arabinofuranosyl units of the arabinan of the cell wall of mycobacteria (Scherman *et al.*, 1995). These may be obtained from the hexose monophosphate shunt (HMP) pathway in *M. leprae*, which accounts for nearly 10% of metabolized glucose in *M. leprae*. A complete set of HMP genes exists in *M. leprae*.

Fatty Acid and Mycolic Acid Synthesis

Homologous genes for the two types of fatty acid synthases generally found in mycobacteria are present in *M. leprae*: the multifunctional single peptide fatty acid synthase FAS-I (*fas*) (Fernandes and Kolattukudy, 1996) and the disassociated multienzyme FAS-II (*fabD, acpM, kasA, kasB, accD6, inhA* and *fabG*) (Mdluli *et al.*, 1998). The C_{14}-C_{26} acyl-CoA products synthesized by FAS-I are used for the fatty acids of plasma membrane lipids, the α-chain of mycolic acids and also for extension by the FAS-II synthase to C_{50-56} to form the meromycolate branch of the mycolic acids. There is no ortholog in *M. leprae* for the FabH β-ketoacyl-acyl carrier protein (ACP) synthase III of *M. tuberculosis* (mtFabH) which has been shown to catalyze

Figure 1. Gene reduction for energy yielding pathways in *M. leprae*: The pathways of glycolysis, tricarboxylic acid cycle, glyoxylate cycle, fermentation (anaerobic respiration), hexosemonophosphate shunt, β-oxidation of fatty acids and oxidative phosphorylation are shown schematically. Dotted arrows indicate multiple steps. The pseudogenes of *M. leprae* are underlined and *M. tuberculosis* genes that are absent in *M. leprae* are in bold letters. The reaction catalyzed by phosphoenolpyruvate carboxylase encoded by the *ppc* gene is present in *M. leprae* and absent in *M. tuberculosis* and may represent a major route for the entry of oxaloacetate into the tricarboxylic aid cycle. The hexosemonophosphate pathway provides NADPH (reducing power) for fatty acid synthesis. The uptake and oxidation of glucose, glucose-6-P, 6-phosphogluconate, glycerol and palmitic acid have been shown by metabolic labeling of whole *M. leprae* cells and are indicated by double-lined arrows. Note that the pool size of acetyl-CoA for tricarboxylic acid cycle and fatty acid synthesis is diminished due to a number of factors: *pta, lpdA, pdhA, pdhB,* and *pdhC* are pseudogenes; the presence of only one lipase for release of fatty acids from host lipids; and the absence of the fatty acid transporter FadD6. Fermentation of glucose to lactate, ethanol and formate are also impacted by gene loss. The subsequent build up of NADH is compounded by the lack of one (Ndh1) of the two NADH oxidases involved in oxidative phosphorylation.

the condensation of C_{14}-CoA and malonyl-ACP to generate a C_{16}-ACP primer for further elongation with KasA and KasB ketoacyl synthases of the FAS-II complex (Choi *et al.*, 2000, Schaeffer *et al.*, 2001). The *de novo* fatty acid synthesis measured as malonyl-CoA incorporation was very insignificant in cell free assays (Wheeler *et al.*, 1990) and also in whole cells maintained on external carbon sources such as citrate, glucose and pyruvate (Wheeler and Ratledge., 1988b). The latter may be attributed to the diminished production of acetyl-CoA from these substrates as described previously (Figure 1) and the lack of FabH.

De novo Purine Synthesis

Enzymes for *de novo* synthesis of the purine ring were not detected in extracts of *M. leprae*, whereas uptake of purines (hypoxanthine, guanine, adenine) or nucleosides (adenosine) and their incorporation into nucleic acids was shown in whole bacteria (Wheeler, 1987). However, according to the genome, all genes for the biosynthesis of purines appear to be present in *M. leprae* except for the non-essential *purU* and *purT*, which are present in *M. tuberculosis*. The PurT has dual activities: acetate kinase (Acetate + ATP → Acetyl-phosphate + ADP) and 5' phosphoribosylglycinamide transformylase [β-glycinamide ribonucleotide (GAR) + ATP + Formate → Formyl-β-GAR] in *E. coli* (Nagy *et al.*, 1993). The formate for the transformylase comes from the reaction catalyzed by PurU (Formyl-tetrahydrofolate → Formate + Tetrahydrofolate). In *E. coli*, another gene product, PurN, also catalyzes the formation of Formyl-β-GAR, but may not depend on the formate produced by PurU, and only *purT, purN* double mutants are auxotrophs for purine (Nygaard and Smith, 1993). The role of the alternative PurU-PurT enzymes is postulated to be for the regulation of formate/tetrahydrofolate levels and is controlled by the levels of glycine and methionine (Nagy *et al.*, 1995). Why is *M. leprae* an auxotroph for purine, and can we avail of this in achieving axenic growth? The enzymes for pyrimidine synthesis were detected in cell free assays (Wheeler, 1989b), and the complete complement of genes for their synthesis are present in the genome.

Amino Acids, Metals, and Cofactor Requirements

From the genome, it appears that biosynthesis of amino acids is intact, except for methionine (*metC*) (Cole *et al.*, 2001). There are many pseudogenes for transport of cations (magnesium, nickel, mercury, ammonium, ferrous and ferric ions and potassium) and anions (arsenate, sulfate and phosphate). The pathway for sulfur metabolism (*cysT, cysW, cysA*) is defective, and, therefore, a source of sulfur is best supplied in any attempts at cultivation (Cole *et al.*, 2001). Iron uptake is, of course, very important for mycobacteria. However, *M. leprae* does not have the genes for the synthesis of mycobactin or its uptake of iron as seen in *M. tuberculosis*, though a mycobactin-like compound has been reported in *M. leprae* (Dhople and Osborne, 1988b). It has been shown that heterologous mycobacterial exochelins can bind Fe^{+3}, and be taken up by *M. leprae* through passive diffusion (Hall *et al.*, 1983). Bacterioferritin, an iron binding protein seen in *M. leprae* (Pessolani *et al.*, 1994) is encoded by *bfrA*, while the redundant *bfrB* gene, present in *M. tuberculosis*, is a pseudogene in *M. leprae* (Cole *et al.*, 2001). In terms of cofactors such as FAD and NAD, the pathways for their synthesis are apparently intact in *M. leprae*; however, purine deficiency may affect levels of NAD and FAD. Apparently, all of the genes required for the biosynthesis of cobinamide are pseudogenes, and, hence, cobalamine would have to be supplied.

Osmoregulation

Given the fact that *M. leprae* has defective transport systems and osmoregulation genes (*proX, proV, proZ*), it is likely that maintenance of osmolarity may be a factor in its slow growth rate. Growth of *M. leprae* in the presence of trehalose (a good osmoprotectant) should be considered, since enzymes for its synthesis are limited in *M. leprae* (the TreY-TreZ pathway, one of three pathways, is inactive) (DeSmet *et al.*, 2000). Other reagents such as sucrose, glycerol, maltose, cellobiose, gentiobiose, turanose, palatinose, and polyethylene glycol may also be used, but they may also serve as carbon sources.

Detoxification

The presence of many pseudogenes of peroxidases (*katG, hpx, ahpE, ahpD*), epoxides (*ephF*) and cytochrome P450 enzymes may have a detrimental effect on *M. leprae* growth. Addition of peroxidases to the medium could be beneficial.

Quorum Sensing

In several bacteria including mycobacteria, the growth rate depends on cell density. Intercellular communication via molecules released into the medium can control cell division. Whether such pathways are operative in *M. leprae* is not known.

Reagents for Diagnosis and Vaccines

An urgent need in leprosy research is better diagnosis of infected persons, especially in endemic areas in order to curtail the trend of increasing incidence of leprosy or new case detection. Present "laboratory" methods are neither specific nor sensitive for leprosy diagnosis, but have some value for disease classification. These "laboratory" methods, which rely on more sensitive T-cell mediated rather than B-cell mediated responses, are based on reactivity to crude extracts of *M. tuberculosis* (purified protein derivative, PPD) (Snider, 1982) or leprosy-nodule derived antigens including lepromin (Krotoski, 1993) or the Rees and Convit *M. leprae* soluble antigen (MLSA) from armadillo derived *M. leprae* (Anonymous, 1985). New skin test antigens from armadillo-derived *M. leprae* prepared by the removal of lipids and lipoglycans from the MLSA (MLSA-LAM) and cell wall (MLCwA) have been shown to be sensitive and have some specificity in *M. leprae*-sensitized guinea pigs (Brennan and TerLouw, 1998). To improve specificity, antigenic proteins in MLSA-LAM are being size-fractionated by preparative electroelution followed by another round of testing in sensitized guinea pigs (Brennan, 2000). In addition, the membrane fraction of *M. leprae* contains unique antigens. Such methods for detecting *M. leprae*-specific B and T cell antigens using native proteins can now be augmented by the availability of the genome and its analysis with suitable bioinformatic tools. Beyond the identification of proteins unique to *M. leprae*, it is necessary to apply certain criteria to obtain good B or T cell antigenic candidates. In the design adopted by Dockrell *et al.* (Dockrell *et al.*, 2000), the *M. leprae* genome sequence was scanned for T cell epitopes using the algorithms FINDPATTERN and Epimer. In this method, "promiscuous T-cell epitopes" that can potentially bind more than one MHC type were identified and tested for the ability to induce lymphocyte proliferation and release of IFN-γ only in tuberculoid leprosy patients or their close contacts and not those infected with *M. tuberculosis* or healthy individuals in non-endemic areas. By using a pool of peptides, a kit can then be developed for diagnosis in individuals with different genetic backgrounds. The same antigens can also be tested as vaccines. The vaccine may be delivered as a cocktail of synthetic peptides or as a DNA vaccine (where individual peptides are encoded in separate plasmids or as a fusion of several peptides in one plasmid). Such genome driven vaccine development has also been applied for the diagnosis of HIV-1 infection and tuberculosis (De Groot *et al.*, 2001).

Secreted or surface exposed proteins are usually good B and T cell antigens. The culture filtrate proteins of *M. tuberculosis* have been valuable for the discovery of antigens and enzymes that have a direct application in the development of vaccines and diagnostic tools (Covert *et al.*, 2001; Samanich *et al.*, 1998; Sonnenberg and Belisle, 1997). Nowadays, genomes can readily be analyzed for the full spectrum of secreted proteins produced by a bacterium. Localization of proteins can be predicted using a combination of algorithms that look for signal peptides, transmembrane helices, lipoprotein signatures and host-cell binding domains. For serodiagnosis and/or vaccine development for leprosy, the labor intensive, but effective and extensive method applied to *Meningococcus* may be useful. From the analysis of the genome of the serogroup B virulent *Neisseria meningitidis*, 570 putative CDSs for surface exposed proteins were found (Pizza *et al.*, 2000). The genes were cloned and expressed in *E. coli* as His-tagged or glutathione S-transferase (GST) fusion proteins. Recombinant proteins were successfully obtained for 350 CDSs and antibodies were raised in mice. Of these, 85 proteins elicited antibodies active in one of three assays: detection of protein on the surface of a panel of different naturally occurring strains by fluorescent activated cell sorting (FACS); detection of protein at a low antibody dilution in ELISA; and induction of complement mediated bactericidal activity. Seven of these proteins met all the three criteria and were deemed to be useful vaccine candidates. Four of them were found to be putative lipoproteins. In the *M. leprae* genome, there are several genes for putative lipoproteins.

The value of secreted proteins for diagnostics and vaccines is highlighted in the study of the 6-kDa ESAT-6 (early secreted antigenic target) and the 10-kDa L_{45} homologous protein (also called CFP-10) family of proteins in tuberculosis and leprosy (Berthet *et al.*, 1998). The genes for ESAT-6 (*esx*) and CFP-10 (*lhp*)-like proteins are found in tandem as part of a cluster of conserved genes in several species of mycobacteria (*M. avium, M. paratuberculosis, M. smegmatis*, and *M. bovis*) and the related high G+C Gram-positive *Corynebacterium diptheriae* and *Streptomyces coelicolor* (Gey Van Pittius *et al.*, 2001). Five clusters, resulting from gene duplication events, are found in *M. tuberculosis* (denoted as regions 1-5). In *M. leprae*, only two are intact (regions 1 and 3) (Figure 2A); two others have many

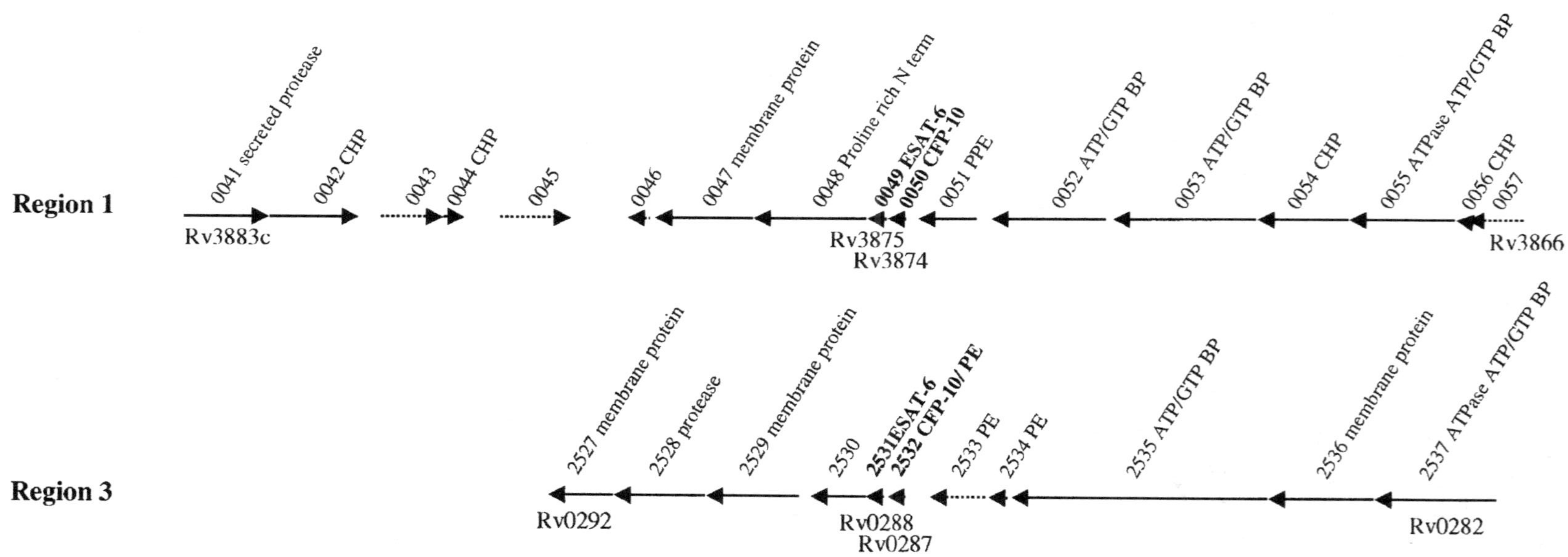

Figure 2A. The organization of the ESAT-6 clusters in *M. leprae*. The ESAT-6 clusters designated as region 1 and 3 by Gey van Pittius *et al.* (2001) are shown. Each cluster includes the genes encoding ESAT-6 and CFP-10 homologs and the proteins of their putative secretion apparatus. The gene numbers and their putative functions, shown above the arrows are as per the annotations in the *M. leprae* genome database. The dotted arrows represent pseudogenes in *M. leprae*. The homologous *M. tuberculosis* genes at the ends of the regions are indicated below the arrows. Abbreviations: CHP: conserved hypothetical protein; PE: protein belonging to the proline-glutamic acid family; PPE: protein belonging to the proline-proline-glutamic acid family; ATP/GTP BP: ATP/GTP binding protein.

```
                   10        20        30        40        50        60        70        80        90
ML0049     MIQAWHFPALQGAVNELQGSQSRIDALLEQCQESLTKLQSSWHGSGNESYSSVQRRFNQNTEGINHALGDLVQAINHSAETMQQTEAGVMSMFTG
             : :.: ....:.. .::. . : .::.. ..::::: ..: :::.:.:.::..... ..  .:.:: .:..:.....: .::..: .::.
Rv3875    MTEQQWNFAGIEAAASAIQGNVTSIHSLLDEGKQSLTKLAAAWGGSGSEAYQGVQQKWDATATELNNALQNLARTISEAGQAMASTEGNVTGMFA
                  10        20        30        40        50        60        70        80        90
36% identity in 91 aa overlap

                  10        20          30             40        50        60        70        80        90
ML2531    MTQIMYNYPAMLDHAGNMSACAGALQGVG--IDIAAEQ-----AALQACWGGDTGISYQAWQVQWNQATEEMVRAYHAMANTHQNNTLAMLTRDQAEAAKWGG
                       :    ... .. :::    ::   ::      . ::. : :. . ::.. : ..:: :: .     ::...  :   . .  : .:.::.
ML0049            MIQAWHFPALQGAVNELQGSQSRIDALLEQCQESLTKLQSSWHGSGNESYSSVQRRFNQNTEGI---NHALGDLVQAINHSAETMQQTEAGVMSMFTG
                          10        20        30        40        50        60           70        80        90
30% identity in 85 aa overlap

                  10        20        30        40        50        60        70        80        90
ML2532    MSLLDVHIPQLVASESAFAAKAALMRSQINQAECEAISAQAFHQGESSAAFQSAHAQFVTAAEKIN-ALLDIAQQHLGEAAETYVATDATAASTYTTGL
                :.: : .. . . .. . . . ..: .    . :.  .: .. ...:.. .:   .: :: :: :..:  ....:::.  :.: . : .:
ML0049     MIQAWHFPALQGAVNELQGSQSRIDALLEQCQESLTKLQSSWHGSGNESYSSVQRRFNQNTEGINHALGDLVQA-INHSAETMQQTEAGVMSMFTG
                   10        20        30        40        50        60        70        80        90
24% identity in 90 aa overlap

                  10        20        30        40        50        60        70        80        90       100
ML0363    MGEVARGSGVVLTAPTKSNTPLAGKSTARPARFNAALEELRSQIAPLQQRWIRAAVVVYPVEQLKWHQAATALNEILVDLGNVIRAGAEEVANTDRRSGGVWAG
                                   .:..: ::. . ... ::. :  ..   :   : ...: . ..:. : :: ..:  .:: . .:.        :
ML0049                MIQAWHFPALQGAVNELQGSQSRIDALLEQCQESLTKLQSSWHGSGNESYSSVQRRFNQNTEGINHALGDLVQAINHSAETMQQTEAGVMSMFTG
                              10        20        30        40        50        60        70        80        90
27% identity in 65 aa overlap
```

Figure 2B. The organization of the ESAT-6 clusters in *M. leprae*. The amino acid alignments between the ESAT-6 of *M. leprae* (ML0049) and *M. tuberculosis* (Rv3875) and each of the putative *M. leprae* homologs ML2531, ML2532 and ML0363. Similarity between amino acids is represented by one dot and identity with two dots. The dominant B cell and T cell epitopes of *M. leprae* ESAT-6 in mice are indicated with bold and underlined fonts respectively (Spencer *et al.*, 2002).

pseudogenes (regions 2 and 5) and in one, only the ESAT-6 like gene (ML0363) remains (region 4). The genes ML2531 and ML2532 in region 3 of *M. leprae* have been assigned as the ESAT-6 and CFP-10 homologs, respectively, even though they are both more homologous to the prototype ESAT-6 gene ML0049 than with the CFP-10 gene ML0050 (Figure 2B). The other genes in the clusters encode membrane associated proteins, such as putative ABC transporters and subtilisin-like proteases, thought to be involved in the secretion and processing of ESAT-6 and CFP-10. If true, it is expected that the product of the gene ML0363 will not be secreted. The ESAT-6 gene (ML0049) of region 1 of *M. leprae* has been cloned and expressed (Spencer *et al.*, 2002). T cell hybridomas and monoclonal antibodies were generated against the recombinant protein and tested against a panel of overlapping peptides. It was found that despite the overall homology to the *M. tuberculosis* ortholog, T cell and antibody reactivity was not cross-reactive in mice, indicating its promise as a skin test antigen and vaccine. The *M. tuberculosis* ESAT-6 has been applied towards diagnosis of tuberculosis since its counterpart is deleted from the genome of the vaccine strain *M. bovis BCG* (van Pinxteren *et al.*, 2000). We suggest that it would be worthwhile to study the other ESAT-6 and CFP-10 homologs of *M. leprae* and *M. tuberculosis* (Table 1).

Other proteins that may be good antigens are members of the PE, PPE and PE-PGRS family of proteins. There are very few intact PE and PPE encoding genes in the *M. leprae* genome compared to *M. tuberculosis* and none for PE-PGRS (Table 1). There are three sets of tandem PE and PPE genes (ML0410, ML0411; ML0538, ML0539; ML1183, ML1182). Three other PPE genes are by themselves (ML0051, ML1991, ML1828). The other putative PE genes are ML1053 and ML2534. Some of the PE and PPE genes are located within the ESAT-6 regions (ML0051 in region 1 and ML2534 in region 3, Figure 2A and Table 1). Interestingly, the ML2532 gene annotated as a PE gene in the *M. leprae* genome database has been designated as the

Table 1. Genomic arrangement of the ESAT-6, CFP-10, PE and PPE family of genes in *M. tuberculosis* H37Rv and identification of corresponding loci in *M. leprae*. Homologs of the genes encoding the *M. tuberculosis* H37Rv ESAT-6 and Lhp (CFP-10) antigens were identified in the *M. leprae* genome and their genomic arrangement was compared with that in the *M. tuberculosis* genome. In the table, the abbreviation N/O (no ortholog) is used when a homologous gene was not found in a similar genetic context. The homologs of CFP-10 and ESAT-6 genes that are located downstream of PE and PPE genes are shown in bold, whereas those that are not are in regular font. The PE and PPE genes that are not linked to ESAT-6 like genes are shown in italics; the underlined genes represent pseudogenes. The numbers in the column 'ESAT-6 cluster region' are as defined by Gey van Pittius *et al.*, 2001.

Table 1. Genomic arrangement of the ESAT-6, CFP-10, PE and PPE family of genes in *M. leprae*

M. tuberculosis	*M. leprae*	% Identity	% Similarity	ESAT-6 cluster region
Rv3872 (PE)	**N/O**			1
Rv3873 (PPE)	**ML0051**	52	64	1
Rv3874 (Lhp)	**ML0050**	40	60	1
Rv3875 (ESAT-6)	**ML0049**	36	65	1
Rv3890c	**ML0034**			2
Rv3891c	N/O			2
Rv3892c (PPE)	N/O			2
Rv3893c (PE)	N/O			2
Rv0285 (PE)	**ML2534 (PE)**	73	84	3
Rv0286 (PPE)	**ML2533 (PE)**			3
Rv0287	**ML2532**	76	85	3
Rv0288	**ML2531**	70	85	3
Rv3444c	ML0363	71	81	4
Rv3445c	N/O			4
Rv1790 (PPE)	N/O			5
Rv1791(PE)	N/O			5
Rv1792	N/O			5
Rv1793	N/O			5
Rv1037c	N/O			
Rv1038c	N/O			
Rv1039c (PPE)	N/O			
Rv1040c (PE)	N/O			
Rv1195 (PE)	**ML1053**	48	67	
Rv1196 (PPE)	**ML1054**			
Rv1197	**ML1055**	58	77	
Rv1198	**ML1056**	64	78	
Rv3619c	N/O			
Rv3620c	N/O			
Rv3621c (PPE)	N/O			
Rv3622c (PE)	N/O			
Rv3019c	N/O			
Rv3020c (PE)	N/O			
Rv3021c (PPE)	N/O			
Rv3022c (PPE)	N/O			
N/O	**ML1180**			
N/O	**ML1181**			
N/O	**ML1182 (PPE)**			
N/O	**ML1183 (PE)**			
Rv2346c	N/O			
Rv2347c	N/O			
Rv3904c	N/O			
Rv3905c	N/O			
Rv1386 (PE)	*ML0538 (PE)*	71	76	
Rv1387 (PPE)	*ML0539 (PPE)*	67	80	
Rv2107 (PE)	*ML0410 (PE)*	44	61	
Rv2108 (PPE)	*ML0411(PPE)*	41	55	
Rv0096 (PPE)	*ML1991 (PPE)*	69	78	
Rv0256c (PPE)	*ML1828 (PPE)*	59	67	

CFP-10 homolog of region 3 as described above. Recently, the PE domain of a *M. tuberculosis* PE-PGRS protein (Rv1818c) was demonstrated to be a T cell antigen, while the PGRS domain is a B cell antigen. In the presence of both domains, the native PE-PGRS protein masks the T cell antigenicity (Delogu and Brennan, 2001).

In the case of vaccine development for leprosy, it is useful to take into consideration the disease states manifested according to the immune status of the individuals for the definition of correlates of protection. Lima *et al.* (Lima *et al.*, 2000) suggested that good vaccine candidates should elicit an immune response as seen in exposed but leprosy-resistant contacts. It was demonstrated that in leprosy-resistant individuals, exposure to *M. leprae* leads to the emergence of a protective response, characterized by secretion of IFN-γ at low levels in association with moderate TNF-α/IL-10 ratios and IL-5 production resembling a Th0 type response. The vaccination of recently exposed contacts with BCG also elicits this type of cytokine profile. Another model for "protection" is based on the role of IL-18 as a highly pleiotropic, pro-Th1 cytokine that enhances both innate and acquired host immune responses. IL-18 induces macrophages to produce IFN-γ, TNF-α and nitric oxide (NO), protective against a broad range of pathogens including *Mycobacterium* spp. (Sugawara, 2000). Vaccine candidates (recombinant proteins that contain putative MHC binding motifs or cruder subcellular fractions) can be assayed for the protective cytokine profile induced in *in vitro* stimulated PBMC from contact individuals before and after BCG vaccination and also from tuberculoid leprosy patients.

Identification of Drug Targets in the Cell Envelope

The complex cell envelope of mycobacteria contributes to its structural integrity, acts as a permeability barrier to toxic molecules (Brennan and Nikaido, 1995), allows specific interaction with the host during infection, and is responsible for the evasion of host immune responses and pathogenicity (Chatterjee and Khoo, 1998). Thus, study of the biosynthesis of the individual components and their assembly is informative for discovering drug candidates (Crick and Brennan, 2000). Much has been learned regarding the structure and composition in the case of several species of mycobacteria, including *M. leprae* (Brennan and Nikaido, 1995). A common aspect is that beyond the inner membrane is a cell wall core comprised of a covalently linked complex of three components: peptidoglycan, a branched polysaccharide called arabinogalactan, and the long chain branched lipids known as mycolic acids. In addition, the phosphatidylinositol mannosides (PIMs), and the related lipomannan (LM) and lipoarabinomannan (LAM) are inserted non-covalently in the inner membrane and are also found in the cell wall and on the surface of the cell wall proper. The outer surface may also contain non-covalently linked molecules such as phthiocerol dimycocerosoic acids (PDIMs) and/or its glycosylated-phenolic derivatives; trehalose based lipids such as trehalose monomycolate (TMM) and trehalose dimycolate (TDM), sulfolipids, and lipid-linked oligosaccharides (LOSs); glycopeptidolipids (GPLs); and glucans and mannans. The presence of lipoproteins, glycoproteins and porins in the cell envelope has been described in some species.

The distinguishing features of the *M. leprae* cell wall core include the substitution of L-ala residues with glycine in the peptidoglycan, shorter α-branch (C_{20} vs. $C_{24\text{-}26}$) in the mycolic acids and the absence of methoxy functions on the β-chain of mycolic acids (Draper *et al.*, 1987). The structure of the arabinogalactan appears to be similar to that of *M. tuberculosis*, except that the galactan portion is shorter (Daffe *et al.*, 1993). In *M. leprae*, sulfolipids, glycolipids, TDM, and LOSs are absent, but large amounts of the highly antigenic phenolic glycolipid I (PGL-I) and LAM are present (Hunter and Brennan, 1981). PGL-I is a glycosylated phenolpthiocerol dimycocerosoic acid whose glycosyl moiety is recognized by patient sera (Brennan *et al.*, 1994) and is the basis for the dip-stick diagnostic for multibacillary leprosy (Buhrer-Sekula *et al.*, 2000). The glycosyl moiety is also specifically recognized by the G domain of the α2 chain of laminin-2 on Schwann cells (Rambukkana, 2001). This discovery is significant in understanding the specific tropism of *M. leprae* for Schwann cells and may be used for the prevention of nerve damage in leprosy. The structural details of *M. leprae* LAM have been recently elucidated (D. Chatterjee, unpublished). It differs from the *M. tuberculosis* H37Rv and H37Ra counterparts in having a shorter α 1→6 linked mannan core but with more α1→2 branching. The arabinan chains attached to the mannan core are also shorter, less branched, and have fewer mannose caps at the non-reducing termini. In addition, there is a significantly higher degree of substitution of the entire structure with succinate groups. All of these distinct properties of LAM could provide interesting epitopes for specific diagnosis of leprosy, particularly where PGL-I fails, as in paucibacillary leprosy. The biochemistry and genetics of the cell wall core synthesis are being elucidated more rapidly since the release of the *M. tuberculosis* genome sequence. In recent reviews, we have pointed out the conservation of genes in *M. leprae* for the synthesis of the cell wall core (Vissa and Brennan, 2001). Several gaps still exist in our knowledge of cell wall synthesis, particularly in the identification of all of the glycosyltransferases responsible for the synthesis of arabinogalactan, LM, and LAM; the enzymes for the synthesis of decaprenyl-phosphoryl-arabinofuranose, the donor of arabinofuranose for synthesis of the arabinans of AG and LAM; the condensation enzyme(s) for the conjugation of the α and β chains in mycolic acid synthesis; the hydroxylase for the modification of the N-acetyl muramic acid of peptidoglycan to N-glycolyl muramic acid; and the transglycosylases and transpeptidases for peptide cross-linking. It is anticipated that comparative genomics in combination with suitable *in vitro* assays will result in the identification of these key enzymes. As anticipated, the genes described for sulfolipid and glycolipid synthesis are absent in *M. leprae*.

With regard to some of the protein components, a 45-kDa glycoprotein (also called MPT32, Apa and FAP) in *M. tuberculosis* has been characterized (Dobos *et al.*, 1995). The gene encoding the counterpart was identified in an *M. leprae* λgt11 expression library screened with pooled leprosy patient serum and was named 43L. The recombinant 43L protein was recognized by antiserum against the total culture filtrate proteins of *M. tuberculosis* and also by polyclonal antibodies against the 45-kDa of *M. tuberculosis* (Wieles *et al.*, 1994). The gene that encodes the 45-kDa protein has also been referred to as *modD* in the *M. tuberculosis* genome as it is closely linked to the *modA, modB* and *modC* genes possibly involved in molybdate uptake; in *M. leprae, modA*, *modB* and *modC* are pseudogenes. The antigenicity and protective efficacy of the 45-kDa protein has been attributed to its glycosylation (Horn *et al.*, 1999). Two other major immunodominant proteins of *M. tuberculosis* are the 19-kDa and the 38-kDa lipoproteins (Vordermeier *et al.*, 1997, Young and Garbe, 1991). The 19-kDa protein has also been shown to be glycosylated (Garbe *et al.*, 1993). The *M. leprae* gene ML1966 that encodes a putative lipoprotein is homologous to the 19-kDa of *M. tuberculosis*. The 38-kDa protein, also called PstS1 or PhoS1 or PAB shares homology with two other proteins PstS2 and PstS3 of *M. tuberculosis*; all three proteins are involved in the uptake of phosphate (Lefevre *et al.*, 1997). In the *M. leprae* genome, two homologous genes ML2192 and ML2095 are present, the latter perhaps being the ortholog of *pstS3* of *M. tuberculosis*. The acyl functions of the 19-kDa lipoproteins have a role in eliciting a host immune response. Several mechanisms have been described. The binding to Toll-like receptors on macrophages and their activation is implicated in killing of phagocytosed *M. tuberculosis* (Thoma-Uszynski *et al.*, 2001). Lipopeptides of 19-kDa that bind to class I MHC and are recognized by cytotoxic T-lymphocytes have been suggested as a mechanism for killing of *M. tuberculosis* (Fonseca *et al.*, 2000). The release of reactive nitric oxide induced by lipids of 19-kDa and LAM by infected macrophages is another possible bactericidal pathway (Chan *et al.*, 2001). A 24.1-kDa *M. leprae* protein with lipid modification motif and homologs in *M. tuberculosis* and *M. bovis* BCG has been identified that is recognized by human T cells (Oftung *et al.*, 1997). Another lipoprotein that plays an opposite role in mycobacterial infections is *sodC* encoded Cu-Zn superoxide dismutase that protects mycobacteria from reactive oxygen species released in macrophages (Piddington *et al.*, 2001). Collectively these data support a role for lipoproteins in immunity. The genome of *M. leprae* encodes at least 18 other putative lipoproteins, and their characterization is thus justified.

Epidemiological Tracking of M. leprae *Strains*

In the field of leprosy, fundamental questions still to be addressed definitively are the source of infectious *M. leprae* and route of transmission in humans. The inability to culture *M. leprae in vitro*, combined with its long incubation period prior to onset of symptoms, and the heterogeneity in the disease states it elicits in humans tend to compound the challenge of tracking the prevalence and incidence of *M. leprae* even in endemic parts of the world. However, the issue needs to be tackled before long to achieve elimination of leprosy, especially because, despite a successful MDT program, prevalence is decreasing but incidence is apparently increasing. In the case of tuberculosis, strain typing based on single nucleotide polymorphism (SNPS), variable number of tandem repeats (VNTR), and mobile element IS6110 have been useful for identification of disease clusters, tracking geographical dissemination, and distinguishing between re-infection and relapse (Mazars *et al.*, 2001; Van Soolingen, 2001). However, the genome of *M. leprae* possesses only three phage-like genes, whereas there are many more in that of *M. tuberculosis*. This may explain the lack of heterogeneity of clinical strains in restriction fragment length polymorphism (RFLP) analysis (Williams *et al.*, 1990). However, Fsihi and Cole discovered that polymorphism in the *polA* locus in *M. leprae* was associated with the repeated elements RLEP-L and RLEP-R that belong to a conserved family called RLEP, and they put forth two models for this observation (Fsihi and Cole, 1995). The first hypothesis is that RLEP-L, *polA*, and RLEP-R resemble a composite transposon and can transpose to an alternative locus. The second hypothesis is that the REP-L and REP-R, elements flanking *polA*, can recombine with other pairs of RLEP repeats sites in the genome. In the *M. leprae* genome, 37 RLEP elements have been identified. Of these 37, the regions containing RLEP copies 4 and 5 or 12 and 13 are good candidates for homologous recombination with the RLEP flanking the *polA* locus, and with each other. Polymorphism arising from such recombination can be studied with specific primers. Other loci for allelic diversity have also been discovered. The *rpoT* locus contains a GACATC sequence that is repeated three or four times in clinical strains (Matsuoka *et al.*, 2000b). The TTC triplet sequence, found upstream of a putative sugar transporter pseudogene is more variable and 10 to 37 copies have been detected in clinical strains (Shin *et al.*, 2000). There are other such loci where one, two or more base pairs are repeated, but diversity at these markers has not yet been reported. For a unique fingerprint, the allelic diversity should be very high, and, therefore, several loci may have to be analyzed simultaneously. More recently a novel method using mass spectrometry and cleavable tags attached to allele specific oligionucleotides (primers) has been developed (Kokoris *et al.*, 2000) and tested for *M. tuberculosis* by Qiagen Corporation and TIGR (http://www.tigr.org/new/press_release_3-14-01.shtml) and may be a model for *M. leprae* in the future.

Prospects

The direction and the pace at which research is progressing in the study of *M. tuberculosis* since the genome was deciphered are very encouraging. We envision that despite the daunting challenges unique to leprosy research and the decline of interest in leprosy research parallel to the decline in prevalence, the application of the *M. leprae* genome to the remaining research questions will be equally profound. Clearly, the greatest need from the research community, as defined by the epidemiologists and leprosy control workers, is new tools for the sensitive diagnosis of leprosy, particularly at the population level, in order to accurately gauge the effects of intervention measures, such as MDT, on non-symptomatic leprosy, and thus to provide evidence on routes of transmission and reservoirs of infections. Whether there is a need for a vaccine is debatable, not to mention how it might be implemented in the absence of sensitive diagnostic tools and in the light of low disease prevalence. In combination with classical biochemical and genetic methods, all of the new bioinformatic tools will likely provide leads in the areas of diagnostics and vaccine development. We expect that collaborative and concerted efforts among the few research and clinical institutions still involved in leprosy research will be undertaken to accelerate and optimally utilize the genome information to develop new diagnostics. One such effort may be the cloning and expression of all of the putative genes in a heterologous system. Owing to its limited size and coding capacity, the *M. leprae* genome has been described as a minimal set for a basic mycobacterial system (Cole *et al.*, 2001; Vissa and Brennan, 2001). The biotechnology for high throughput processes of cloning, expression and purification of recombinant proteins is becoming available and has been applied towards all (~6000) *S. cerevisiae* and several hundred human genes (http://www.resgen.com/products/genestorm.php3). The cloning of all of the *E. coli* CDSs is underway (http://www.genome.wisc.edu/functional genomics/functional.html). Likewise, efforts to implement a molecular approach to the epidemiological tracking of *M. leprae* infections will require a concerted effort between the microbial geneticists and the few remaining leprosy research centers, those that can generate the bacillus and propagate it in mouse footpads or at least provide skin biopsy materials. The *in vitro* cultivation of *M. leprae* will probably still be difficult to achieve, because the defects in the *M. leprae* genome are in all areas of metabolism and the factors that control cell division/cell cycle are not known. Thus, the scientific, intellectual, and research challenges in leprosy, although not necessarily new, can be addressed from a new utterly rational perspective, which lends great excitement to the field. And the funding prospects are good with a new infusion of support from the National Institutes of Health, ILEP, other members of GAEL, and some of the European Community structures.

Acknowledgements

Research conducted in the authors' laboratory was supported by NIH, NIAID, DMID Contract NO1 AI-55262 and Grant AI-47197 from the National Institute of Allergy and Infectious Diseases, NIH, and by the Heiser Program for Research in Leprosy and Tuberculosis, New York City. We thank Marilyn Hein, Hee Jin Kim, and Marcus Lambert for assistance with preparation of the manuscript.

References

Andersson, J.O., and Andersson, S.G. 2001. Pseudogenes, junk DNA, and the dynamics of *Rickettsia* genomes. Mol. Biol. Evol. 18: 829-839.

Andersson, S.G., Zomorodipour, A., Andersson, J.O., Sicheritz-Ponten, T., Alsmark, U.C., Podowski, R.M., Naslund, A.K., Eriksson, A.S., Winkler, H.H., and Kurland, C.G. 1998. The genome sequence of *Rickettsia prowazekii* and the origin of mitochondria. Nature. 396: 133-140.

Anonymous. 1985. Vaccination trials against leprosy.TDR/IMMLEP/EPD/ 85.3.Geneva pp. 7-8.

Bapat, C.V. 1989. Cultivation of Mycobacterium leprae: a new approach. Int. J. Lepr. 57: 874-879.

Berthet, F.X., Rasmussen, P.B., Rosenkrands, I., Andersen, P., and Gicquel, B. 1998. A *Mycobacterium tuberculosis* operon encoding ESAT-6 and a novel low-molecular-mass culture filtrate protein (CFP-10). Microbiology. 144: 3195-3203.

Brennan, P.J. 2000. Skin test development in leprosy: progress with first-generation skin test antigens, and an approach to the second generation. Lepr. Rev. 71 Suppl: S50-54.

Brennan, P.J., Chatterjee, D., Fujiwara, T., and Cho, S.N. 1994. Leprosy-specific neoglycoconjugates: synthesis and application to serodiagnosis of leprosy. Methods Enzymol. 242: 27-37.

Brennan, P.J., and Nikaido, H. 1995. The envelope of mycobacteria. Annu. Rev. Biochem. 64: 29-63.

Brennan, P.J., and Terlouw, S.M. 1998. New skin test antigens for the diagnosis of leprosy. Transactions of the 19th International Leprosy Congress. Int. J. Lepr. 66: 96.

Buhrer-Sekula, S., Sarno, E.N., Oskam, L., Koop, S., Wichers, I., Nery, J.A., Vieira, L.M., de Matos, H.J., Faber, W.R., and Klatser, P.R. 2000. Use of ML dipstick as a tool to classify leprosy patients. Int. J. Lepr. 68: 456-463.

Chan, E.D., Chan, J., and Schluger, N.W. 2001. What is the role of nitric oxide in murine and human host defense against tuberculosis? Current knowledge. Am. J. Respir. Cell Mol. Biol. 25: 606-612.

Chatterjee, D., and Khoo, K.H. 1998. Mycobacterial lipoarabinomannan: an extraordinary lipoheteroglycan with profound physiological effects. Glycobiology. 8: 113-120.

Choi, K.H., Kremer, L., Besra, G.S., and Rock, C.O. 2000. Identification and substrate specificity of beta -ketoacyl (acyl carrier protein) synthase III (mtFabH) from *Mycobacterium tuberculosis*. J. Biol. Chem. 275: 28201-28207.

Cole, S.T., Brosch, R., Parkhill, J., Garnier, T., Churcher, C., Harris, D., Gordon, S.V., Eiglmeier, K., Gas, S., Barry, C.E., 3rd, Tekaia, F., Badcock, K., Basham, D., Brown, D., Chillingworth, T., Connor, R., Davies, R., Devlin, K., Feltwell, T., Gentles, S., Hamlin, N., Holroyd, S., Hornsby, T., Jagels, K., Krogh, A., McLean, J., Moule, S., Murphy, L., Oliver, K., Osborne, J., Quail, M.A., Rajandream, M.-A., Rogers, J., Rutter, S., Seeger, K., Skelton, J., Squares, S., Squares, R., Sulston, J.E., Taylor, K., Whitehead, S., and Barrell, B.G. 1998. Deciphering the biology of *Mycobacterium tuberculosis* from the complete genome sequence. Nature 393: 537-544.

Cole, S.T., Eiglmeier, K., Parkhill, J., James, K.D., Thomson, N.R., Wheeler, P.R., Honore, N., Garnier, T., Churcher, C., Harris, D., Mungall, K., Basham, D., Brown, D., Chillingworth, T., Connor, R., Davies, R.M., Devlin, K., Duthoy, S., Feltwell, T., Fraser, A., Hamlin, N., Holroyd, S., Hornsby, T., Jagels, K., Lacroix, C., Maclean, J., Moule, S., Murphy, L., Oliver, K., Quail, M. A., Rajandream, M.A., Rutherford, K.M., Rutter, S., Seeger, K., Simon, S., Simmonds, M., Skelton, J., Squares, R., Squares, S., Stevens, K., Taylor, K., Whitehead, S., Woodward, J.R., and Barrell, B.G. 2001. Massive gene decay in the leprosy bacillus. Nature. 409: 1007-1011.

Covert, B.A., Spencer, J.S., Orme, I.M., and Belisle, J.T. 2001. The application of proteomics in defining the T cell antigens of *Mycobacterium tuberculosis*. Proteomics. 1: 574-586.

Crick, D.C., and Brennan, P.J. 2000. Antituberculosis drug research. Curr. Opin. Anti-infect. Invest. Drugs. 2: 154-163.

Daffe, M., McNeil, M., and Brennan, P.J. 1993. Major structural features of the cell wall arabinogalactans of *Mycobacterium*, *Rhodococcus*, and *Nocardia* spp. Carbohydr. Res. 249: 383-398.

De Groot, A.S., Bosma, A., Chinai, N., Frost, J., Jesdale, B.M., Gonzalez, M.A., Martin, W., and Saint-Aubin, C. 2001. From genome to vaccine: in silico predictions, ex vivo verification. Vaccine. 19: 4385-4395.

De Smet, K.A., Weston, A., Brown, I.N., Young, D.B., and Robertson, B.D. 2000. Three pathways for trehalose biosynthesis in mycobacteria. Microbiology. 146: 199-208.

Delogu, G., and Brennan, M.J. 2001. Comparative immune response to PE and PE_PGRS antigens of *Mycobacterium tuberculosis*. Infect. Immun. 69: 5606-5611.

Dhople, A.M., Green, K.J., and Osborne, L.J. 1988a. Limited in vitro multiplication of *Mycobacterium leprae*. Ann. Inst. Pasteur. Microbiol. 139: 213-223.

Dhople, A.M., and Osborne, L.J. 1988b. Presence of mycobactin-like substance in *Mycobacterium leprae*. Indian J. Lepr. 60: 348-359.

Dobos, K.M., Swiderek, K., Khoo, K.H., Brennan, P.J., and Belisle, J.T. 1995. Evidence for glycosylation sites on the 45-kilodalton glycoprotein of *Mycobacterium tuberculosis*. Infect. Immun. 63: 2846-2853.

Dockrell, H.M., Brahmbhatt, S., Robertson, B.D., Britton, S., Fruth, U., Gebre, N., Hunegnaw, M., Hussain, R., Manandhar, R., Murillo, L., Pessolani, M.C., Roche, P., Salgado, J.L., Sampaio, E., Shahid, F., Thole, J.E., and Young, D.B. 2000. A postgenomic approach to identification of *Mycobacterium leprae*-specific peptides as T-cell reagents. Infect. Immun. 68: 5846-5855.

Draper, P., Kandler, O., and Darbre, A. 1987. Peptidoglycan and arabinogalactan of *Mycobacterium leprae*. J. Gen. Microbiol. 133: 1187-1194.

Fernandes, N.D., and Kolattukudy, P.E. 1996. Cloning, sequencing and characterization of a fatty acid synthase-encoding gene from *Mycobacterium tuberculosis* var. bovis BCG. Gene. 170: 95-99.

Fonseca, D.P., Joosten, D., Snippe, H., and Verheul, A.F. 2000. Evaluation of T-cell responses to peptides and lipopeptides with MHC class I binding motifs derived from the amino acid sequence of the 19-kDa lipoprotein of *Mycobacterium tuberculosis*. Mol. Immunol. 37: 413-422.

Fsihi, H., and Cole, S.T. 1995. The *Mycobacterium leprae* genome: systematic sequence analysis identifies key catabolic enzymes, ATP-dependent transport systems and a novel polA locus associated with genomic variability. Mol. Microbiol. 16: 909-919.

Ganapati, R., and Revankar, C.R. 1989. Clinical aspects of leprosy. In: The biology of the mycobacteria Vol. 3. C. Ratledge, J. Stanford, and J.M. Grange, eds., Academic Press Limited, London. p. 327-358.

Garbe, T., Harris, D., Vordermeier, M., Lathigra, R., Ivanyi, J., and Young, D. 1993. Expression of the *Mycobacterium tuberculosis* 19-kilodalton antigen in *Mycobacterium smegmatis*: immunological analysis and evidence of glycosylation. Infect. Immun. 61: 260-267.

Gey Van Pittius, N.C., Gamieldien, J., Hide, W., Brown, G.D., Siezen, R.J., and Beyers, A..D. 2001. The ESAT-6 gene cluster of *Mycobacterium tuberculosis* and other high G+C gram-positive bacteria. Genome Biol. 2: research 0044.1-0044.18.

Hall, R.M., Wheeler, P.R., and Ratledge, C. 1983. Exochelin-mediated iron uptake into *Mycobacterium leprae*. Int. J. Lepr. 51: 490-494.

Honer Zu Bentrup, K., Miczak, A., Swenson, D.L., and Russell, D.G. 1999. Characterization of activity and expression of isocitrate lyase in *Mycobacterium avium* and *Mycobacterium tuberculosis*. J. Bacteriol. 181: 7161-7167.

Honore, N., and Cole, S.T. 1993. Molecular basis of rifampin resistance in *Mycobacterium leprae*. Antimicrob. Agents Chemother. 37: 414-418.

Horn, C., Namane, A., Pescher, P., Riviere, M., Romain, F., Puzo, G., Barzu, O., and Marchal, G. 1999. Decreased capacity of recombinant 45/47-kDa molecules (Apa) of *Mycobacterium tuberculosis* to stimulate T lymphocyte responses related to changes in their mannosylation pattern. J. Biol. Chem. 274: 32023-32030.

Hunter, S.W., and Brennan, P.J. 1981. A novel phenolic glycolipid from *Mycobacterium leprae* possibly involved in immunogenicity and pathogenicity. J. Bacteriol. 147: 728-735.

Kai, M., Matsuoka, M., Nakata, N., Maeda, S., Gidoh, M., Maeda, Y.,

Hashimoto, K., Kobayashi, K., and Kashiwabara, Y. 1999. Diaminodiphenylsulfone resistance of *Mycobacterium leprae* due to mutations in the dihydropteroate synthase gene. FEMS Microbiol. Lett. 177: 231-235.

Kokoris, M., Dix, K., Moynihan, K., Mathis, J., Erwin, B., Grass, P., Hines, B., and Duesterhoeft, A. 2000. High-throughput SNP genotyping with the Masscode system. Mol. Diagn. 5: 329-340.

Krahenbuhl, J., and Adams, L.B. 2000. Exploitation of gene knockout mice models to study the pathogenesis of leprosy. Lepr Rev. 71 Suppl: S170-175.

Krotoski, W.A., Mroczkowski, T.F., Rea, T.H., Almodovar, P.I., Clements, B.C., Neimes, R.E., Kahkonen, M.K., Job, C.K., and Hastings, R.C. 1993. Lepromin skin testing in the classification of Hansen's disease in the United States. Am. J. Med. Sci. 305: 18-24.

Lee, B.Y., and Horwitz, M.A. 1995. Identification of macrophage and stress-induced proteins of *Mycobacterium tuberculosis*. J. Clin. Invest. 96: 245-249.

Lefevre, P., Braibant, M., de Wit, L., Kalai, M., Roeper, D., Grotzinger, J., Delville, J.P., Peirs, P., Ooms, J., Huygen, K., and Content, J. 1997. Three different putative phosphate transport receptors are encoded by the *Mycobacterium tuberculosis* genome and are present at the surface of *Mycobacterium bovis* BCG. J. Bacteriol. 179: 2900-2906.

Levy, L. 1976. Studies of the mouse foot pad technique for cultivation of *Mycobacterium leprae*. 3. Doubling time during logarithmic multiplication. Lepr. Rev. 47: 103-106.

Lima, M.C., Pereira, G.M., Rumjanek, F.D., Gomes, H.M., Duppre, N., Sampaio, E.P., Alvim, I. M., Nery, J.A., Sarno, E.N., and Pessolani, M.C. 2000. Immunological cytokine correlates of protective immunity and pathogenesis in leprosy. Scand. J. Immunol. 51: 419-428.

Marques, M.A., Chitale, S., Brennan, P.J., and Pessolani, M.C. 1998. Mapping and identification of the major cell wall-associated components of *Mycobacterium leprae*. Infect. Immun. 66: 2625-2631.

Matsuoka, M., Kashiwabara, Y., and Namisato, M. 2000a. A *Mycobacterium leprae* isolate resistant to dapsone, rifampin, ofloxacin and sparfloxacin. Int. J. Lepr. 68: 452-455.

Matsuoka, M., Maeda, S., Kai, M., Nakata, N., Chae, G.T., Gillis, T.P., Kobayashi, K., Izumi, S., and Kashiwabara, Y. 2000b. *Mycobacterium leprae* typing by genomic diversity and global distribution of genotypes. Int. J. Lepr. 68: 121-128.

Mazars, E., Lesjean, S., Banuls, A. L., Gilbert, M., Vincent, V., Gicquel, B., Tibayrenc, M., Locht, C., and Supply, P. 2001. High-resolution minisatellitebased typing as a portable approach to global analysis of *Mycobacterium tuberculosis* molecular epidemiology. Proc. Natl. Acad. Sci. USA. 98: 1901-1906.

Mdluli, K., Slayden, R.A., Zhu, Y., Ramaswamy, S., Pan, X., Mead, D., Crane, D.D., Musser, J.M., and Barry, C.E., 3rd. 1998. Inhibition of a *Mycobacterium tuberculosis* beta-ketoacyl ACP synthase by isoniazid. Science. 280: 1607-1610.

Monahan, I.M., Betts, J., Banerjee, D.K., and Butcher, P.D. 2001. Differential expression of mycobacterial proteins following phagocytosis by macrophages. Microbiology. 147: 459-471.

Nagy, P.L., Marolewski, A., Benkovic, S.J., and Zalkin, H. 1995. Formyltetrahydrofolate hydrolase, a regulatory enzyme that functions to balance pools of tetrahydrofolate and one-carbon tetrahydrofolate adducts in *Escherichia coli*. J. Bacteriol. 177: 1292-1298.

Nagy, P.L., McCorkle, G.M., and Zalkin, H. 1993. purU, a source of formate for purT-dependent phosphoribosyl-N-formylglycinamide synthesis. J. Bacteriol. 175: 7066-7073.

Nakamura, M., and Matsuoka, M. 2001. Limited ATP generation in cells of *Mycobacterium leprae* Thai-53 strain in enriched Kirchner liquid medium containing adenosine. Int. J. Lepr. 69: 13-20.

Nygaard, P., and Smith, J.M. 1993. Evidence for a novel glycinamide ribonucleotide transformylase in *Escherichia coli*. J. Bacteriol. 175: 3591-3597.

Oftung, F., Wiker, H.G., Deggerdal, A., and Mustafa, A.S. 1997. A novel mycobacterial antigen relevant to cellular immunity belongs to a family of secreted lipoproteins. Scand. J. Immunol. 46: 445-451.

Parkhill, J., Wren, B.W., Thomson, N.R., Titball, R.W., Holden, M.T., Prentice, M.B., Sebaihia, M., James, K.D., Churcher, C., Mungall, K.L., Baker, S., Basham, D., Bentley, S.D., Brooks, K., Cerdeno-Tarraga, A.M., Chillingworth, T., Cronin, A., Davies, R.M., Davis, P., Dougan, G., Feltwell, T., Hamlin, N., Holroyd, S., Jagels, K., Karlyshev, A.V., Leather, S., Moule, S., Oyston, P.C., Quail, M., Rutherford, K., Simmonds, M., Skelton, J., Stevens, K., Whitehead, S., and Barrell, B.G. 2001. Genome sequence of *Yersinia pestis*, the causative agent of plague. Nature. 413: 523-527.

Pessolani, M.C., Smith, D.R., Rivoire, B., McCormick, J., Hefta, S.A., Cole, S.T., and Brennan, P.J. 1994. Purification, characterization, gene sequence, and significance of a bacterioferritin from *Mycobacterium leprae*. J. Exp. Med. 180: 319-327.

Piddington, D.L., Fang, F.C., Laessig, T., Cooper, A.M., Orme, I.M., and Buchmeier, N.A. 2001. Cu,Zn superoxide dismutase of *Mycobacterium tuberculosis* contributes to survival in activated macrophages that are generating an oxidative burst. Infect. Immun. 69: 4980-4987.

Pizza, M., Scarlato, V., Masignani, V., Giuliani, M.M., Arico, B., Comanducci, M., Jennings, G. T., Baldi, L., Bartolini, E., Capecchi, B., Galeotti, C.L., Luzzi, E., Manetti, R., Marchetti, E., Mora, M., Nuti, S., Ratti, G., Santini, L., Savino, S., Scarselli, M., Storni, E., Zuo, P., Broeker, M., Hundt, E., Knapp, B., Blair, E., Mason, T., Tettelin, H., Hood, D.W., Jeffries, A.C., Saunders, N.J., Granoff, D.M., Venter, J.C., Moxon, E.R., Grandi, G., and Rappuoli, R. 2000. Identification of vaccine candidates against serogroup B meningococcus by whole-genome sequencing. Science. 287: 1816-1820.

Prabhakaran, K. 1967. Oxidation of 3,4-dihydroxyphenylalanine (DOPA) by *Mycobacterium leprae*. Int. J. Lepr. 35: 42-51.

Ragno, S., Romano, M., Howell, S., Pappin, D.J., Jenner, P.J., and Colston, M.J. 2001. Changes in gene expression in macrophages infected with *Mycobacterium tuberculosis*: a combined transcriptomic and proteomic approach. Immunology. 104: 99-108.

Rambukkana, A. 2001. Molecular basis for the peripheral nerve predilection of *Mycobacterium leprae*. Curr. Opin. Microbiol. 4: 21-27.

Ridley, D.S., and Jopling, W.H. 1966. Classification of leprosy according to immunity. A five-group system. Int. J. Lepr. 34: 255-273.

Samanich, K.M., Belisle, J.T., Sonnenberg, M.G., Keen, M.A., Zolla-Pazner, S., and Laal, S. 1998. Delineation of human antibody responses to culture filtrate antigens of *Mycobacterium tuberculosis*. J. Infect. Dis. 178: 1534-1538.

Sanchez, J.C., Golaz, O., Frutiger, S., Schaller, D., Appel, R.D., Bairoch, A., Hughes, G.J., and Hochstrasser, D.F. 1996. The yeast SWISS-2DPAGE database. Electrophoresis. 17: 556-565.

Sassetti, C.M., Boyd, D.H., and Rubin, E.J. 2001. Comprehensive identification of conditionally essential genes in mycobacteria. Proc. Natl. Acad. Sci. U S A. 98: 12712-12717.

Schaeffer, M.L., Agnihotri, G., Volker, C., Kallender, H., Brennan, P.J., and Lonsdale, J.T. 2001. Purification and biochemical characterization of the *Mycobacterium tuberculosis* beta-ketoacyl-acyl carrier protein synthases KasA and KasB. J. Biol. Chem. 276: 47029-47037.

Scherman, M., Weston, A., Duncan, K., Whittington, A., Upton, R., Deng, L., Comber, R., Friedrich, J.D., and McNeil, M. 1995. Biosynthetic origin of mycobacterial cell wall arabinosyl residues. J. Bacteriol. 177: 7125-7130.

Shannon, E.J., Harris, E.B., Haile-Mariam, H.S., Guebre-Xavier, M., and Frommel, D. 1992. Competency of human-derived *Mycobacterium leprae* to use palmitic acid in the synthesis of phenolic glycolipid-I and phthiocerol dimycocerosate and to release CO_2 in axenic culture. Lepr. Rev. 63: 101-107.

Shepard, C.C. 1965. Temperature optimum of *Mycobacterium leprae* in mice. J. Bacteriol. 90: 1271-1275.

Sherman, D.R., Voskuil, M., Schnappinger, D., Liao, R., Harrell, M.I., and Schoolnik, G.K. 2001. Regulation of the *Mycobacterium tuberculosis* hypoxic response gene encoding alpha-crystallin. Proc. Natl. Acad. Sci. USA. 98: 7534-7539.

Shin, Y.C., Lee, H., Walsh, G.P., Kim, J.D., and Cho, S.N. 2000. Variable numbers of TTC repeats in *Mycobacterium leprae* DNA from leprosy patients and use in strain differentiation. J. Clin. Microbiol. 38: 4535-4538.

Smida, J., Kazda, J., and Stackebrandt, E. 1988. Molecular-genetic evidence for the relationship of *Mycobacterium leprae* to slow growing pathogenic mycobacteria. Int. J. Lepr. 56: 449-454.

Snider, D.E., Jr. 1982. The tuberculin skin test. Am. Rev. Respir. Dis. 125: 108-118.

Sonnenberg, M.G., and Belisle, J.T. 1997. Definition of *Mycobacterium tuberculosis* culture filtrate proteins by two-dimensional polyacrylamide gel electrophoresis, N-terminal amino acid sequencing, and electrospray mass spectrometry. Infect. Immun. 65: 4515-4524.

Spencer, J.S., Marques, M.A., Lima, M.C., Junqueira-Kipnis, A.P., Gregory, B.C., Truman, R.W., and Brennan, P.J. 2002. Antigenic specificity of the *Mycobacterium leprae* homologue of ESAT-6. Infect. Immun. 70: 1010-1013.

Sugawara, I. 2000. Interleukin-18 (IL-18) and infectious diseases, with special emphasis on diseases induced by intracellular pathogens. Microbes. Infect. 2: 1257-1263.

Thoma-Uszynski, S., Stenger, S., Takeuchi, O., Ochoa, M.T., Engele, M., Sieling, P.A., Barnes, P.F., Rollinghoff, M., Bolcskei, P.L., Wagner, M., Akira, S., Norgard, M.V., Belisle, J.T., Godowski, P.J., Bloom, B.R., and Modlin, R.L. 2001. Induction of direct antimicrobial activity through mammalian toll-like receptors. Science. 291: 1544-1547.

Tonella, L., Walsh, B.J., Sanchez, J.C., Ou, K., Wilkins, M.R., Tyler, M., Frutiger, S., Gooley, A. A., Pescaru, I., Appel, R.D., Yan, J.X., Bairoch, A., Hoogland, C., Morch, F.S., Hughes, G.J., Williams, K.L., and Hochstrasser, D.F. 1998. '98 *Escherichia coli* SWISS-2DPAGE database update. Electrophoresis. 19: 1960-1971.

Truman, R.W., and Krahenbuhl, J.L. 2001. Viable *M. leprae* as a research reagent. Int. J. Lepr. 69: 1-12.

van Pinxteren, L.A., Ravn, P., Agger, E.M., Pollock, J., and Andersen, P. 2000. Diagnosis of tuberculosis based on the two specific antigens ESAT-6 and CFP10. Clin. Diagn. Lab. Immunol. 7: 155-160.

Van Soolingen, D. 2001. Molecular epidemiology of tuberculosis and other mycobacterial infections: main methodologies and achievements. J. Intern. Med. 249: 1-26.

Vissa, V.D., and Brennan, P.J. 2001. The genome of *Mycobacterium leprae*: a minimal mycobacterial gene set. Genome Biol. 2: reviews 1023.1-1023.8.

Vordermeier, H.M., Zhu, X., and Harris, D.P. 1997. Induction of CD8+ CTL recognizing mycobacterial peptides. Scand. J. Immunol. 1997. 45: 521-526.

Weber, I., Fritz, C., Ruttkowski, S., Kreft, A., and Bange, F.C. 2000. Anaerobic nitrate reductase (narGHJI) activity of *Mycobacterium bovis* BCG in vitro and its contribution to virulence in immunodeficient mice. Mol. Microbiol. 35: 1017-1025.

Wheeler, P.R. 1984. Metabolism in *Mycobacterium leprae*: its relation to other research on *M. leprae* and to aspects of metabolism in other mycobacteria and intracellular parasites. Int. J. Lepr. 52: 208-230.

Wheeler, P.R. 1987. Biosynthesis and scavenging of purines by pathogenic mycobacteria including *Mycobacterium leprae*. J. Gen. Microbiol. 133 : 2999-3011.

Wheeler, P.R. 1989a. Biosynthetic pathways in *Mycobacterium leprae*. Acta Leprol. 7 Suppl 1: 21-24.

Wheeler, P.R. 1989b. Pyrimidine biosynthesis in *Mycobacterium leprae* and other intracellular mycobacteria. Acta Leprol. 7 Suppl 1: 33-35.

Wheeler, P.R., Bulmer, K., and Ratledge, C. 1990. Enzymes for biosynthesis de novo and elongation of fatty acids in mycobacteria grown in host cells: is *Mycobacterium leprae* competent in fatty acid biosynthesis? J. Gen. Microbiol. 136 : 211-217.

Wheeler, P.R., and Ratledge, C. 1988a. Metabolism in *Mycobacterium leprae*, *M. tuberculosis* and other pathogenic mycobacteria. Br. Med. Bull. 44: 547-561.

Wheeler, P.R., and Ratledge, C. 1988b. Use of carbon sources for lipid biosynthesis in *Mycobacterium leprae*: a comparison with other pathogenic mycobacteria. J. Gen. Microbiol. 134: 2111-2121.

Wieles, B., van Agterveld, M., Janson, A., Clark-Curtiss, J., Rinke de Wit, T., Harboe, M., and Thole, J. 1994. Characterization of a *Mycobacterium leprae* antigen related to the secreted *Mycobacterium tuberculosis* protein MPT32. Infect. Immun. 62: 252-258.

Williams, D.L., Gillis, T.P., and Portaels, F. 1990. Geographically distinct isolates of *Mycobacterium leprae* exhibit no genotypic diversity by restriction fragment-length polymorphism analysis. Mol. Microbiol. 4: 1653-1659.

Williams, D.L., Spring, L., Harris, E., Roche, P., and Gillis, T.P. 2000. Dihydropteroate synthase of *Mycobacterium leprae* and dapsone resistance.

Antimicrob. Agents Chemother. 44: 1530-1537.

Winder, F. 1982. Mode of action of the antimycobacterial agents and the associated aspects of the molecular biology of the mycobacteria. In: The biology of the mycobacteria. C. Ratledge and J. Stanford, eds., Academic press, London. p. 353-438

World Health Organization Study Group. 1982. Chemotherapy of leprosy for control programmes. WHO Technical Report Series No 675. World Health Organization, Geneva.

World Health Organization. 2002. Leprosy-Global situation. Weekly Epidemiol. Rec. 77: 1-8.

Young, D.B., and Garbe, T.R. 1991. Lipoprotein antigens of *Mycobacterium tuberculosis*. Res. Microbiol. 142: 5-65.

From: *Genomics of GC-Rich Gram-Positive Bacteria*

6

Genomics as a Tool for Identifying Secreted Proteins in Bacteria

Yanis Ben Amor and Maria Laura Gennaro

Abstract

Bacteria have evolved at least four pathways to transport proteins out of the cytoplasm. Proteins secreted by one of these pathways, the type II system, are characterized by the presence of a cleavable, NH_2-terminal signal peptide. All other pathways lack an obvious marker for secretion. The sequence information gained by the analysis of bacterial genomes and of the proteins they encode should accelerate the identification of common domains/motifs in proteins targeted by, or constituting the transport machinery of, a given secretion pathway. Since genes involved in secretion by type I, III and IV systems tend to be found in clusters, finding substrates of secretion should help identify the corresponding transporters and vice versa. Three families of well-characterized secreted proteins of *Mycobacterium tuberculosis* were analyzed for relationships between gene location and gene regulation with protein sequence, function, or subcellular location. While it is still not possible to identify a secreted protein based on sequence information alone, there is little doubt that the ever-expanding knowledge of gene sequence, location and context on genomes, together with information on protein structure and function, will help develop rules to predict the fate of proteins in terms of their secretion.

Introductiom

To perform the breadth of functions required for bacterial survival and spread, proteins need to be partitioned to one of the subcellular compartments of the bacterial cell, or, to be released into the extracellular environment. In Gram-positive bacteria, proteins that are translocated across the cell membrane may remain surface-associated or be released into the external milieu. In Gram-negative bacteria, because of the presence of two membranes in the cell envelope, proteins transported out of the cytoplasm have the additional options of being associated with either of two membranes or of being released into the periplasmic space. Most of our current knowledge of bacterial secretion systems has been derived from studies with Gram-negative organisms, primarily because these bacteria have been historically more amenable to genetic enquiry.

Four major pathways of protein secretion have been identified in eubacteria (Salmond *et al.,* 1993; Danese *et al.,* 1998; Buchanan, 2001). The best studied is the *sec*-dependent, general export pathway that targets proteins carrying NH_2-terminal, cleavable signal peptides in both Gram-negative and Gram-positive bacteria (Pugsley, 1993; Danese *et al.*, 1998). More recently, additional pathways of protein secretion have been described that do not require the classical, cleavable signal peptide. Two such pathways are the type I and the type III secretion systems (Binet *et al.*, 1997; Hueck, 1998). They are typically referred to as signal-sequence independent pathways, although it is becoming increasingly clear that they recognize signals for targeting proteins for export. A fourth pathway, the type IV secretion system, includes a number of heterogeneous systems, some that are type-III-like and others that utilize the *sec*-dependent machinery for inner membrane crossing. The ever expanding availability of bacterial genomes and the use of comparative genomics should help investigators determine which are the protein (or even mRNA, see below) sequences recognized by each type of protein secretion machinery. However, it is not currently understood why certain proteins are secreted via one pathway and not another.

Despite the tremendous advance in our understanding of the mechanisms of protein secretion in Gram-negative and Gram-positive bacteria, the nature of the systems used by bacteria characterized as having “atypical” cell envelopes remains largely undefined. In particular, very little is known about the secretion systems of *Mycobacterium tuberculosis*, even though a key role of secreted proteins in the interaction of this pathogen with its host has been long recognized (Andersen *et al.,* 1994). With the publication of the genome sequence of *M. tuberculosis* (Cole *et al.*, 1998) several questions arise: can genomics enable us to identify the secretion mechanisms operating in *M. tuberculosis*? Can protein sequence/structure analysis forecast "secretability" in this organism? Can the context of a gene in the chromosome be used as a "sign" for a given gene product to be a secreted protein?

In this chapter we first review the key properties and the genetic organization of the four major protein secretion pathways of bacteria. Each of these systems has been described in depth in excellent reviews (Table 1). Then we describe protein secretion mechanisms operating in *M. tuberculosis*. Finally, we utilize examples of known secreted proteins of *M. tuberculosis* to examine whether protein sequence, function, and gene location can be used to identify a secreted protein.

Pathways of Protein Secretion in Bacteria

Signal-Sequence-Independent Mechanisms

The Type I System

The type I pathway (Figure 1) is the simplest of the secretion systems described for Gram-negative bacteria, since proteins cross both membranes in one step without formation of a periplasmic intermediate. The central feature of the type I secretion apparatus is an inner-membrane bound, ATP-dependent protein translocator that contains two hydrophobic membrane-spanning domains and two hydrophylic nucleotide-binding sequences on the cytoplasmic side of the membrane (Quentin *et al.*, 1999). Proteins in this family are referred to as ABC (**A**TP-**B**inding **C**assette) proteins (Higgins, 1992). In many bacteria, these domains are synthesized as separate proteins encoded by independent genes that are often organized in operon(s) or at least clustered in a DNA region (Felmlee *et al.*, 1985; Linton *et al.*, 1998). A second component is a largely periplasmic protein that is anchored to the inner membrane by either a lipid anchor or by a single, putative transmembrane α helix (Paulsen *et al.*, 1997). It also has a large hydrophilic domain located in the periplasm and a COOH-terminal sequence that presumably interacts with the outer membrane. These proteins are classified as members of the membrane fusion protein (MFP) family on the basis of their membrane topology (Dinh *et al.*, 1994). The third component of the type I secretion pathway is the outer membrane channel, which, at least in the case of the TolC protein (Buchanan, 2001), has been shown by crystal

Table 1. Pathways of protein secretion in Eubacteria

Type[a]	Steps[b]	Components[c]	Specificity	Signal sequence	References
I	1	ABC transporter Membrane Fusion Protein TolC protein	Specific Specific Non Specific	COOH-terminal COOH-terminal COOH-terminal	Binet *et al.* (1997) Braibant *et al*, (2000) Buchanan (2001) Thanabalu *et al.* (1998)
II	2	Sec proteins ~ 14 proteins	Non specific Non specific	Cleavable NH_2-terminal Variable	Anderson (1999) Cornelis (2000) Danese *et al.* (1998) Pugsley (1993)
III	1	20 proteins or more	Non specific	NH_2-terminal 5' end mRNA	Aldridge *et al.* (2001) Anderson *et al.* (1999) Hueck (1998) Portnoy *et al.* (1981)
IV	2	Sec proteins Variable	Non specific -	Cleavable NH_2-terminal Variable	Burns (1999) Christie *et al.* (2001)

(a) type of secretion pathway
(b) number of steps involved in the protein secretion process (1, direct secretion into the external milieu; 2, formation of a periplasmic intermediate)
(c) number of proteins involved in the secretion machinery

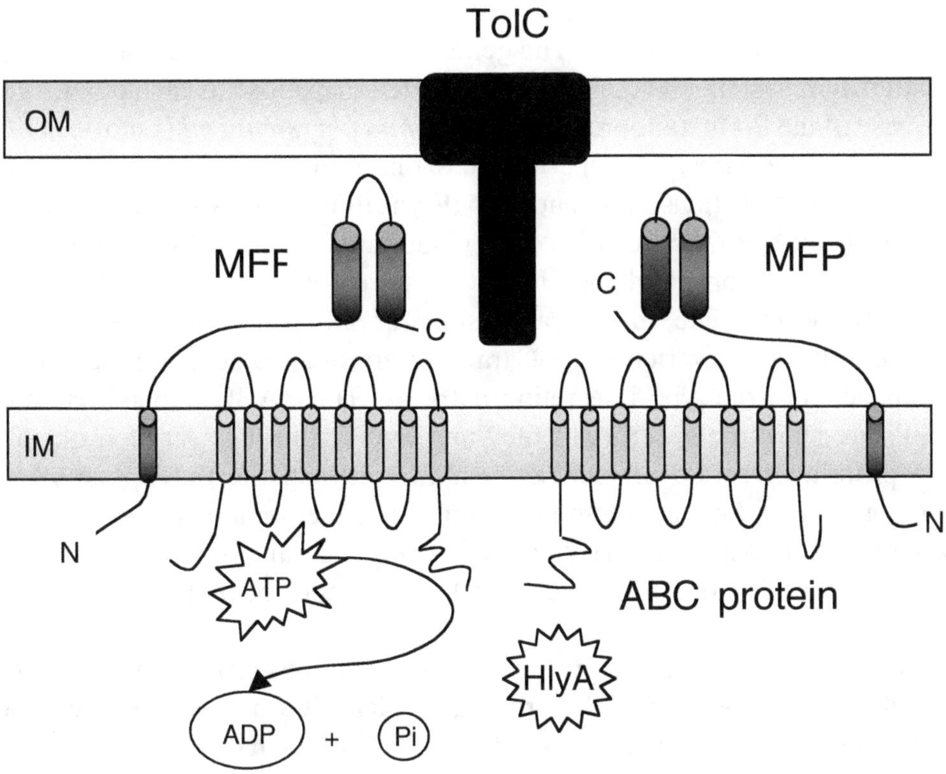

Figure 1. The type I transport system. The type I secretion apparatus, examplified here by the hemolysin (HlyA) export system of *Escherichia coli*, is composed of three parts that interact with each other during the course of secretion to form a continuous channel : (a) an inner-membrane (IM) bound, ATP-dependent protein translocator that is referred to as ATP-Binding Cassette (ABC) protein (shown in light grey), (b) a largely periplasmic protein anchored to the inner membrane and classified as a member of the membrane fusion protein (MFP) family (shown in dark grey) and (c) an outer-membrane (OM) channel, TolC (shown in black). [Modified from (Gentschev *et al.*, 2002)]

diffraction analysis to be exceptionally long (140 Å). The TolC β-barrel channel spans the outer membrane, and its α-helical tunnel extends into the periplasm (Buchanan, 2001). In the presence of substrate, all three export components can be crosslinked (Thanabalu *et al.*, 1998), suggesting that they form a continuous channel that connects the cytoplasmic side of the inner membrane to the extracellular side of the outer membrane. Assembly of the multi-protein secretion apparatus is ordered (Binet *et al.*, 1997): the substrate is recognized by the ABC protein, which then interacts with the MFP. The resulting complex in turn associates with the outer membrane component. The three-component apparatus disassembles following transport of the substrate molecule (Thanabalu *et al.*, 1998).

The type I system mediates secretion of a wide variety of substrates, including toxins, lipases, and proteases. The genes encoding the three components of the transport system and their substrates often map close to each other, as is the case of the Prt metalloproteases of *Erwinia chrysanthemi* (Letoffe *et al.*, 1990). In other cases, the gene for the exoprotein maps close to the genes encoding the ABC transporter and the MFP, as illustrated by the α-hemolysin protein (HlyA) of *Escherichia coli* (Mackman *et al.*, 1986). The outer membrane component of the HlyA transport system is the product of a distantly located gene, *tolC* (Wandersman *et al.*, 1990; Fath *et al.*, 1991). TolC is utilized by a variety of ABC transporters, as are other outer membrane components of the type I secretion pathway. Thus, ABC transporters and MFPs are substrate specific, as they are involved in the secretion of only one protein or group of isoenzymes. In contrast, the outer-membrane component may be coupled to numerous inner-membrane and periplasmic proteins to transport a variety of unrelated substances, including large proteins, peptides, organic solvents, detergents, and antibiotics.

Proteins secreted via the type I pathway contain secretion signals in the COOH-terminal portion of the protein, as identified by gene deletion and gene fusion analyses (Mackman *et al.*, 1986). The COOH-terminal sequence of the *E. coli* HlyA, when fused to a variety of heterologous proteins, is sufficient to promote secretion of the fusion products (Nakano *et al.*, 1992; Blight *et al.*, 1994). The same is true for the COOH-terminal portions of a family of highly homologous metalloproteases of *Er. chrysanthemi* (Delepelaire *et al.*, 1990). In addition, proteins secreted via the type I pathway often contain an additional stretch of repeated Gly-rich sequences located just upstream of the COOH-terminal secretion signal (Letoffe *et al.*, 1992). It has been suggested that these Gly-rich repeats act as internal chaperones to facilitate the formation of a protein conformation optimal for secretion (Binet *et al.*, 1997).

The Type III System

Type III secretion (Figure 2), which is found in many Gram-negative bacterial pathogens, is mediated by 20 or more proteins that enable bacteria adhering to the surface of eukaryotic cells to translocate proteins into the cytosol of an eukaryotic cell. It is likely that type III secretion systems also enable bacteria residing inside vacuoles to inject proteins across the vacuolar membrane into the cytosol. Such proteins, called effectors, alter key functions of the attacked cell often by disturbing cytoskeleton dynamics or by affecting cell signaling cascades. Action of the type III secretion machinery invariably

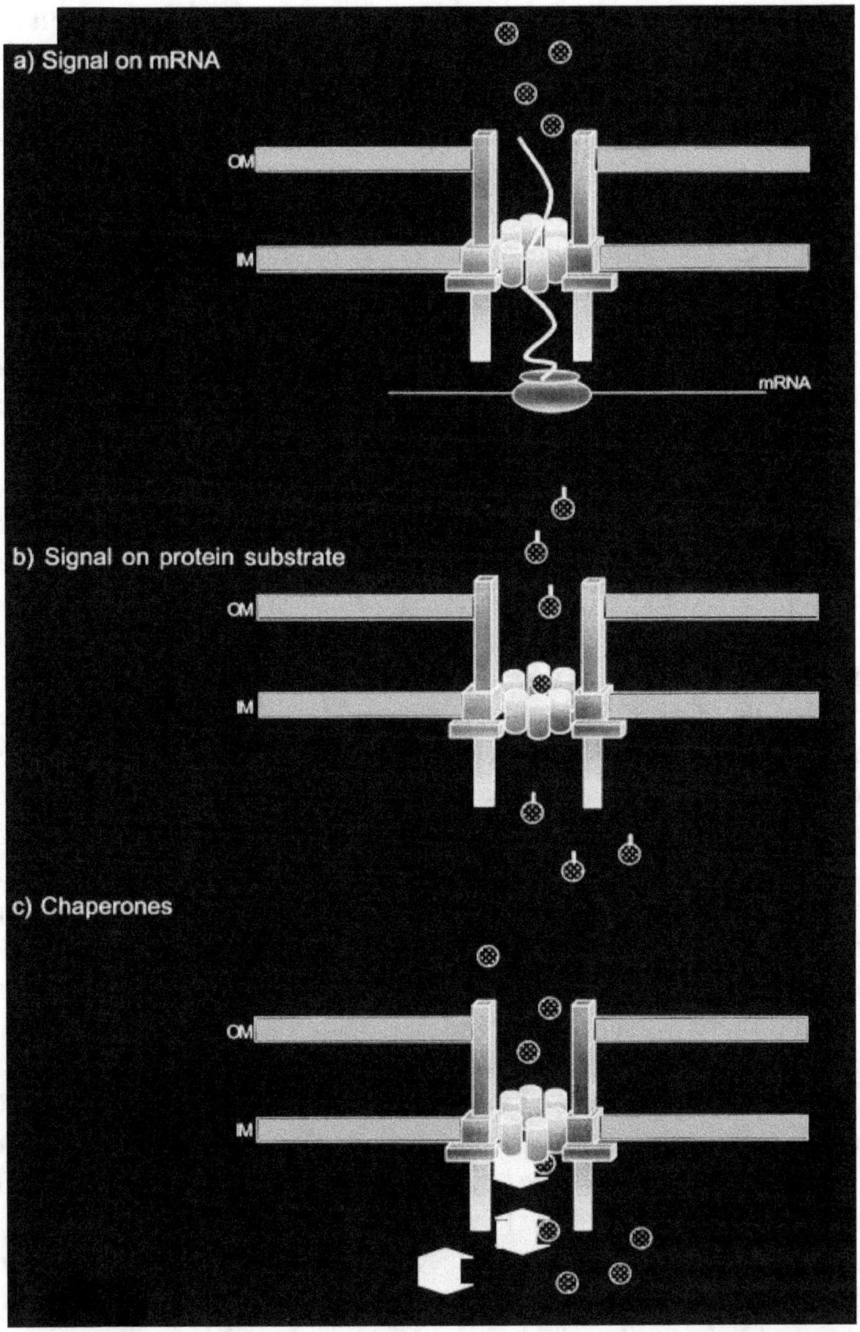

Figure 2. The type III secretion pathway. Three mechanisms can be postulated for the secretion of proteins via the type III system: (a) a mechanism mediated by a secretion signal on the mRNA that leads to coupled translation and secretion of the substrate with the cytoplasmic face of the secretion apparatus; (b) a mechanism mediated by an NH_2-terminal secretion signal on the protein substrate; (c) a mechanism mediated by a cognate chaperone (black symbols) that binds its secretion substrate (dotted circles). The chaperone interacts with the secretion apparatus and delivers the substrate for secretion. [Modified from (Aldridge *et al.*, 2001)]

results in an enhanced ability of the pathogen to enter the target cell and to survive therein (Hueck, 1998). Secretion of the Yop proteins of *Yersinia* spp. (Portnoy *et al.*, 1981), the first family of effector proteins shown to be secreted via the type III secretion pathway, involves 29 gene products. Of these, only a portion has been assigned a function. Studies on the *Salmonella typhimurium* type III secretion system demonstrated that type III secretion occurs through a needle (Kubori *et al.*, 1998), a complex structure comprising a long, hollow conduit that allows translocation of proteins across the bacterial inner and outer membranes in one step. The channel is held in place by multiple, ring-shaped structures associated with the inner and outer membranes. Associated ATPases are presumably required to provide energy for the translocation process.

How substrates are selected for type III secretion remains controversial. Fusion of the reporter enzyme alkaline phosphatase (lacking its secretion signal peptide) to the NH_2-terminal portion of some Yop proteins causes alkaline phophatase to be efficiently secreted into the extracellular medium, suggesting the presence of a secretion signal sequence. However, comparative analysis of the first 50 to 100 NH_2-terminal residues of Yop proteins has failed to reveal a conserved secretion signal peptide. It has been suggested (Anderson *et al.*, 1997; Anderson *et al.*, 1999) that the amino-terminal secretion signal in *Yersinia* is defined by an mRNA sequence, rather than by an amino acid sequence, implying that secretion occurs during translation (Figure 3). However, pulse-chase experiments indicated that secretion by the type III system may occur after translation has been completed, again pointing to a secretion signal in the amino acid sequence (Aldridge *et al.*, 2001). Additionally, many of the proteins secreted via the type III pathway require chaperones. These chaperones directly interact with their substrate either to escort it to the base of the secretion apparatus (Cheng *et al.*, 2000) or to prevent substrate degradation in the cytoplasm. Mutations in the genes coding for the chaperones have been shown to result in aberrant secretion of the cognate secretion substrate (Bennett *et al.*, 2000). These findings imply that, at least in some cases, the presence of a chaperone obviates the need for a signal sequence on the target protein. Recent work on the flagellum export apparatus of *Salmonella*, which is similar to the type III secretion apparatus (Minamino *et al.*, 1999), leads to a model for type III secretion that accommodates mRNA signals, chaperone binding, and NH_2-terminal secretion signal sequences (Aldridge *et al.*, 2001).

It is likely that type III secretion systems have been acquired by horizontal transfer, as they are usually found on plasmids (Portnoy *et al.*, 1981) or within pathogenicity islands (Winstanley *et al.*, 2001), large DNA regions characterized by clustering of virulence factors whose G+C content often differs from that of the rest of the chromosome (Groisman *et al.*, 1997).

Signal-Sequence-Dependent Mechanisms

The Type II System

The type II system is the best understood secretion pathway of bacteria. Proteins translocated by the type II secretion machinery all carry at their NH_2 termini a cleavable secretion signal peptide. While their amino acid sequences vary greatly, secretion signal peptides share important features such as length, presence of basic amino acids at the NH_2 terminus, a hydrophobic core, and a hydrophylic COOH terminal sequence containing the recognition sequence for leader peptidases. Thus, it is possible to identify by amino acid sequence analysis proteins targeted by the type II secretion machinery.

Secretion by the type II pathway occurs in two steps. The first step is shared by Gram-positive and Gram-negative bacteria. It consists of translocation of the target protein across the cytoplasmic membrane and is mediated by the Sec proteins. This step is coupled with the removal from the full-length protein (the preprotein) of the signal peptide by a site-specific endopeptidase and subsequent folding of the resulting mature protein. In Gram-positive organisms, mature proteins may be released into the external environment or anchored to the cell surface either by their NH_2 termini (by lipidation at a cysteine residue of the mature protein that occurs after the removal of the signal peptide) or by their COOH termini [by the sortase-catalyzed anchoring mechanism [reviewed in (Mazmanian *et al.*, 2001)]. In Gram-negative organisms, the proteins secreted into the periplasm by the Sec-dependent machinery are folded and then they either remain in the periplasm or are targeted for translocation across the outer membrane to become outer-membrane proteins or to be released as extracellular enzymes. This second step of the type II secretion pathway is the only translocation system of prokaryotes that uses a folded protein as substrate.

The Sec pathway [reviewed in (Bieker *et al.*, 1990; Murphy *et al.*, 1994; Danese *et al.*, 1998)] can be divided into three distinct, sequential stages:

targeting, translocation and release. During targeting, substrates are guided to exit sites in the membrane. This stage usually requires a secretion-specific chaperone, SecB (Randall *et al.*, 1998), which is responsible for maintaining preproteins in an unfolded state and for directing them to the cytoplasmic membrane. Cytoplasmic chaperones other than SecB have also been associated with protein translocation in *E. coli* (Danese *et al.*, 1998). These include a ribonucleotide complex reminiscent of the mammalian signal-recognition particle (Economou, 1998). At the translocation stage, the protein becomes associated with SecA and thereby with the integral membrane components of the secretion apparatus (SecY, SecE and SecG). SecA has multiple domains that mediate binding and hydrolysis of ATP, binding to the preprotein, to the SecB chaperone, to the integral membrane components of the translocase, and association with the membrane (Economou, 1998). SecA also binds the *secA* mRNA, thereby down-regulating its own synthesis (Schmidt *et al.*, 1989). The ATPase activity of SecA is required for protein-bound SecA to become an integral membrane protein. It is through successive rounds of ATP binding and hydrolysis that SecA binds the preprotein, establishes contact with the secretion apparatus, inserts itself into the inner membrane, and then undergoes several cycles of preprotein release, partial disinsertion from the membrane, reassociation with the preprotein and further translocation. Thus, SecA operates by "pushing" the translocating polypeptide through the membrane in a step-wise fashion. The membrane-bound form of SecA is stabilized by SecD, SecF and YajC. Cleavage of the signal peptide by a leader peptidase occurs during the translocation process. Finally, at the release stage, the translocated, mature polypeptide is released from the Sec apparatus and acquires its native, folded state in the periplasm. SecD may aid in protein folding and/or release from the secretion apparatus (Matsuyama *et al.*, 1993). The folded, periplasmic intermediate may then proceed to the outer membrane for integration or for secretion into the extracellular milieu.

Gram-negative bacteria possess a multi-protein secretion machinery for translocation across the outer membrane [reviewed in (Wandersman, 1993; Russel, 1998; Sandkvist, 2001)]. The mature proteins to be secreted via this pathway are fully folded and often carry the information relevant for secretion in multiple, non-contiguous domains. Secretion signals may be found in the COOH terminus, as in the PehA protein of *Erwinia* spp (Palomaki *et al.*, 1995; Palomaki *et al.*, 1997), or at both termini, as in the *Pseudomonas aeruginosa* exotoxin A (McVay *et al.*, 1995; Lu *et al.*, 1996), or associated with a protein's multimeric structure as in the B-subunit pentamer of cholera toxin (Hirst *et al.*, 1984). Thus, the secretion signal appears to involve a tertiary structure that brings together key residues forming one or several secretion sequences or motifs that are recognized by the secretion proteins. The putative pore of the type II secretion apparatus is composed of a 12- to 14-mer of protein D, a member of the bacterial secretins, which are characterized by a conserved COOH-terminal region embedded in the outer membrane and a variable NH_2-terminal sequence that may be exposed to the periplasm (Genin *et al.*, 1994). An additional outer-membrane protein helps stabilize the pore. The secretion apparatus also comprises proteins associated with the cytoplasmic membrane, one of which, protein E, regulates the secretion process either by phosphorylation or by supplying energy via ATP hydrolysis. Additional inner-membrane proteins are required for secretion, either because they stabilize protein E or because they participate in the relay of information from the cytoplasmic membrane to the pore. The inner-membrane and the outer-membrane components communicate by "bridge" proteins. The last component of the complex type II secretion machinery is constituted by proteins that are homologous to those required for biogenesis of type IV pili. These "pilins" form a pilus-like structure that directly or indirectly supports secretion (Sandkvist, 2001). The similarities shared by these two systems, both at the level of the amino acid sequences of pilins and of the overall supermolecular structure they form, have prompted models for protein translocation across the outer membrane invoking a role for the pilus-like structure in pushing proteins through the secretion pore or in opening the gated pore (Sandkvist, 2001).

The Type IV System

Type IV secretion systems are characterized by a macromolecular transfer machinery whose individual components and overall genetic organization are ancestrally related to conjugal transfer systems found in Gram-negative bacteria (Christie, 2001). Consequently, type IV secretion includes not only some systems devoted to translocation of proteins, such as the pertussin toxin of *Bordetella pertussis* and the CagA protein of *Helicobacter pylori*, but also systems responsible for transfer of nucleoproteins, such as the T-DNA transfer system of *Agrobacterium tumefaciens*.

While type IV secretion systems are all related to conjugal transfer systems, there appears to be no common ancestry. Some type IV secretion systems, such as that of *A. tumefaciens*, are related to the mating pair formation complex that mediates conjugal transfer of DNA, while others, such as the *dot/icm* system of *Legionella pneumophila*, are close homologs of plasmid-borne Tra proteins (Vogel *et al.*, 1999). The *dot/icm* system of *L. pneumophila*, whose effector(s) is unknown, might constitute the prototype of a class of type IV systems thought to contribute to survival of pathogens inside macrophages (Christie, 2001).

Type IV systems also appear to be very heterogeneous with respect to the secretion mechanism they employ (Christie *et al.*, 2000; Christie, 2001). Some resemble a type III secretion system, as they require cytoplasmic chaperone-like proteins, export substrates in a one-step reaction via a trans-envelope channel, and require physical contact with the target cell. Other type IV systems resemble type II secretion systems. For example, the pertussis toxin is translocated by a Sec-dependent mechanism to the periplasm, where it is folded and assembled into its two-subunit, multimeric structure. The type IV pathway then mediates translocation of pertussis toxin across the outer membrane and its release into the extracellular environment. The IgA protease of *Neisseria gonorrhoeae*, whose secretion mechanism is classified as type IV by some authors (Finlay *et al.*, 1997), also utilizes a two-step pathway. The first step, crossing the inner membrane, is mediated by the Sec-dependent machinery. The second step, crossing the outer membrane, is directed by the IgA protease itself, which appears to insert into the outer membrane and to mediate self-release into the medium by autoproteolytic cleavage at three closely spaced sites in its COOH terminal region (Finlay *et al.*, 1997).

Genetic Location of Bacterial Secretion Systems

Some of the properties of the protein secretion systems described above make it possible to predict to what extent genome analyses can help identify secreted proteins in bacteria. For example, for the type I pathway, the genes coding for the secreted substrates are usually adjacent either to the genes for all three components of the transport system or at least to the genes encoding the ABC transporter and the MFP. Thus, identification of ABC transporters, which are characterized by conserved functional domains, should lead to identification of their secretion substrates. Moreover, as the number of known type I secretion substrates increases, comparative sequence and structure analyses of these substrates may help identify the motifs involved in targeting proteins via the type I pathway. The above considerations should also be valid for proteins targeted for secretion by type III pathways, for which even the chemical nature of the secretion signal (the NH_2-terminal amino acid residues and/or the 5' end of the mRNA sequence) remains controversial. The genes encoding type III secretion systems appear to be part of large gene clusters that are presumably acquired by horizontal transfer of extrachromosomal elements or of chromosome regions. Thus, identification of proteins secreted by type III pathways might be aided by analyzing the genome context of their corresponding genes. This task should not be expected to be straightforward, however, because the relative organization of type III secretion genes within a plasmid or a pathogenicity island varies from one system to another (Winstanley *et al.*, 2001), and genes encoding effectors are often scattered throughout the genome, outside of pathogenicity islands (Cornelis *et al.*, 2000). Genomics should undoubtedly help identify type IV secretion systems, which bear similarities in gene organization and sequence to conjugal transfer systems of Gram negative bacteria. Finally, while genetic location might help forecast "secretability" by type I, type III and type IV pathways, it is clear that genes encoding proteins secreted via the type II system do not have proximity with the genes encoding the Sec transport apparatus. For type II secretion, however, the presence of the NH_2-terminal, cleavable signal peptide is a powerful "identifier" of the substrates for secretion.

Protein Secretion in *Mycobacterium tuberculosis*

M. tuberculosis, the causative agent of tuberculosis (TB), is an air-borne pathogen that initiates infection upon ingestion of bacterial cells by alveolar macrophages. Many proteins are found in the extracellular milieu during growth of *M. tuberculosis in vitro* (Sonnenberg *et al.*, 1997). It is not known, however, which proteins are secreted when *M. tuberculosis* is inside the macrophage.

The secreted proteins of *M. tuberculosis* have been studied extensively for the development of TB vaccines and immunodiagnostics. The first indication of a possible association between secreted proteins and protective immunity during TB came from the early observation that live *M. tuberculosis* cells, but not dead bacilli, were capable of inducing protection against challenge in animal models of TB (Bloch *et al.*, 1955; Orme, 1988). Subsequent work by many laboratories established that the supernatant of *M. tuberculosis* cultures (operationally referred to as the culture filtrate) and specific protein components thereof are potent inducers of protective immunity and of immune responses of diagnostic value (Young *et al.*, 1992; Andersen, 1994; Cooper *et al.*, 1995; Andersen, 1997; Collins *et al.*, 2001).

Since the culture filtrate has been the primary source of secreted proteins, a vast effort was devoted to understanding the mechanisms underlying protein release to the external medium during growth of *M. tuberculosis* [reviewed in (Andersen *et al.*, 1994)]. It is generally assumed that proteins found in the culture supernatant at early stages of bacterial growth are those released via specific secretion mechanisms. In contrast, proteins found in the culture medium at late growth stages are those released during bacterial cell autolysis. Based on this concept, actively secreted proteins, irrespective of the mode of their secretion, have been identified as those whose concentration peaked in the culture supernatant when cytosolic markers of cell disruption were still poorly represented (Wiker *et al.*, 1991). However, there are exceptions to this rule. Some proteins that are translocated across the cell envelope may remain associated with the cell surface for some time before being gradually shed into the external environment (Andersen *et al.*, 1994). Moreover, some of the proteins found in the culture filtrate at early stages of growth may also be released via bacterial leakage or through autolysis. Examples are glutamine synthetase and superoxide dismutase, whose abundance in the culture filtrate appears to derive from their high level of expression and extracellular stability rather than to be mediated by a specific secretion mechanism (Tullius *et al.*, 2001).

Despite the interest surrounding extracellular proteins of *M. tuberculosis*, surprisingly little is known about the pathways of protein export in *M. tuberculosis* and in other mycobacteria. One of the challenges in this area of research has been the unique structure and composition of the mycobacterial cell envelope, which is characterized by an arabinogalactan-mycolic acid superpolymer that surrounds and is covalently bound to the peptidoglycan layer (Daffe *et al.,* 1999). While the inner leaflet of a pseudo outer membrane could be constituted by the mycolic acids of the superpolymer, the outer leaflet would be provided by various non-covalently associated phospholipids and glycolipids (Brennan *et al.*, 1994). Thus, even though mycobacteria are formally considered Gram-positive bacteria, their cell wall shows a superficial resemblance to the lipid-rich Gram-negative envelope, but it is notably more waxy and much less fluid (Liu *et al.*, 1996). Thus, some of the mechanisms of protein export found in Gram-negative organisms may also be encountered in mycobacteria.

Sec-Independent Pathways

The presence in the culture filtrate of *M. tuberculosis* of proteins derived from genes that do not encode a NH_2-terminal, cleavable signal peptide that is the hallmark of *sec* dependent secretion has prompted a search for alternative modes of protein secretion. A type I-like secretion system, which utilizes only the ABC transporter component, exists in Gram-positive organisms such as *Enterococcus faecalis* (Gilmore *et al.*, 1990). Indeed ~ 30 putative ABC transporters have been identified in the genome of *M. tuberculosis* on the basis of their amino acid sequences (Braibant *et al.*, 2000). Thus a type I secretion pathway is likely to be present in *M. tuberculosis*. Since genes encoding type I secretion substrates are usually located near those encoding type I secretion factors, genes that may encode secreted substrates have been tentatively identified from among those neighboring some of the putative ABC transporters (Braibant *et al.*, 2000). In contrast with type I secretion system, type III secretion pathways have been identified only in Gram-negative bacteria, and there is no evidence that a type III secretion system is present in *M. tuberculosis*.

The Sec-Dependent Pathway

Substantial evidence supports the presence of a type II, *sec*-dependent pathway in *M. tuberculosis*. Many of the proteins found in the culture filtrate are the products of genes that encode NH_2-terminal, secretion signal peptides (Young *et al.*, 1992; Wiker *et al.*, 1999). Moreover, numerous sequences encoding signal peptides have been identified in gene fusion experiments employing a reporter, leaderless alkaline phosphatase gene (Lim *et al.*, 1995). The genome of *M. tuberculosis* contains homologs of all components of the Sec pathway, except SecB, suggesting that *M. tuberculosis* possesses a *sec*-dependent mechanism to translocate proteins. Interestingly, two homologs of *secA* (*secA1* and *secA2*) have been identified (Cole *et al.*, 1998). The presence of two genes for SecA, unprecedented in other bacterial species, appears to be common among mycobacteria (Braunstein *et al.*, 2001; Cole *et al.*, 2001; Limia *et al.*, 2001). Even though the SecA1 and SecA2 proteins of mycobacteria are homologous to SecA proteins of other organisms, they are only 34% identical to each other. It has been proposed that SecA2 may not be functionally equivalent to SecA1 and that it may assist SecA1 in the translocation of some proteins and operate by itself in the translocation of others (Braunstein *et al.*, 2001).

Identification of Secreted Proteins

In the pre-genome era, the protein composition of the *M. tuberculosis* culture filtrate, which is thought to contain as many as two hundred proteins

(Sonnenberg *et al.*, 1997), was investigated by chromatography, two-dimensional gel electrophoresis, immunological methods, immunochemistry and screening of expression libraries of *M. tuberculosis* DNA with anti-culture-filtrate antisera [see, for example (Andersen *et al.*, 1991; Nagai *et al.*, 1991; Andersen *et al.*, 1992; Amara *et al.*, 1996; Manca *et al.*, 1997; Weldingh *et al.*, 1998; Wiker *et al.*, 1999)]. In addition, secreted or exported proteins were identified by screening libraries of fusions of *M. tuberculosis* DNA to reporter genes encoding enzymes that become active upon translocation across the cell membrane (Lim *et al.*, 1995; Chubb *et al.*, 1998).

The release of the nucleotide sequence of the *M. tuberculosis* genome (Cole *et al.*, 1998) imparted a dramatic turn to TB research. Not only does the genome sequence provide investigators with the possibility to identify proteins that share a common peptide motif (for example, the cleavable, NH_2-terminal signal peptide found in proteins secreted via the type II pathway; see below), but it also fuels "proteome" analyses that couple electrophoretic separation of proteins, mass spectrometry, and matching of amino acid sequences of purified proteins to those deduced from genome analysis (Jungblut, 2001; Mattow *et al.*, 2001). With these powerful tools at hand, the biologist is well placed to combine proteomics with experimental biology to identify the functional organization of the bacterial genome and its preservation among different bacterial species, and to finally decipher "la logique du vivant", the logical explanation of life.

Computer-driven searches for the secreted proteins of *M. tuberculosis* to date have been based on the identification of the NH_2-terminal, cleavable signal peptide required for *sec*-dependent secretion. In a study from our laboratory (Gomez *et al.*, 2000) the *M. tuberculosis* genome was screened for secreted proteins by utilizing four different computer programs. The goal of the work was to identify proteins that were translocated across the cell envelope and released into the external milieu because they carried NH_2-terminal, cleavable signal peptides but lacked membrane-spanning domains and other membrane-anchoring motifs in the putative mature protein (Gomez *et al.*, 2000). Fifty-two such secreted proteins were found. The computer-based identification was tested by reporter gene fusions, and the prediction accuracy was found to be 90%. In another study (Saleh *et al.*, 2001), bacterial genomes were screened for proteins carrying NH_2-terminal, cleavable signal peptides by using a novel program that integrated artificial neural networks. Screening of the *M. tuberculosis* genome with this integrated program predicted that secreted/exported proteins constituted 22% to 26% of the total proteins. In a third study (Schneider, 1999), identification of secreted proteins was based on a more "global" approach in which information for protein targeting was sought along the entire length of the protein. Artificial neural networks were trained on the prediction of the subcellular location of bacterial proteins. It was found that the best accuracy (96% to 97%) was reached for discrimination between extracellular and cytoplasmic proteins. It was also found, however, that grouping of proteins by their predicted subcellular localization remained essentially invariant whether the analysis was conducted by utilizing the full-length sequences or only the 30 most NH_2-terminal amino acid residues of the proteins. Thus, the NH_2-terminal cleavable signal peptide proved again to be the best validated marker for protein secretion. In this study, 23% of the *M. tuberculosis* genome was estimated to code for non-cytoplasmic proteins.

The methods summarized above represent efforts to exploit the *M. tuberculosis* genome sequence to identify secreted and surface-associated proteins by the targeting signal best characterized to date — the NH_2-terminal, cleavable signal peptide used by the type II secretion pathway. And now, whereto? Can we identify proteins secreted by other secretion pathways on the basis of their sequence, structure, function, or of location of their corresponding genes on the genome? In the remaining part of this chapter we will use three families of secreted proteins of *M. tuberculosis* characterized to-date to begin to address the questions delineated above.

Three Families of Secreted Proteins of *Mycobacterium tuberculosis*

Genes Located Together: the ESAT-6 Gene Clusters

Many of the properties of the ESAT-6 antigen of *M. tuberculosis* have placed it at the center of TB immunodiagnostics and vaccine development. ESAT-6 was first described as an immunoreactive protein component of the supernatant of *M. tuberculosis* cultures at early stages of growth (Sorensen *et al.*, 1995). Thus, based on the criteria described in the preceding section, ESAT-6 qualifies as a secreted protein of *M. tuberculosis.* Since its discovery, ESAT-6 has been exhaustively shown to induce strong cell-mediated [(Arend *et al.*, 2000; Pathan *et al.*, 2001; Doherty *et al.*, 2002) to cite a few examples] and humoral (Lyashchenko *et al.*, 1998; Brusasca *et al.*, 2001) immune responses. Recently, ESAT-6 has also been included in subunit and DNA vaccine formulations (Kamath *et al.*, 1999; Weinrich Olsen *et al.*, 2001). The *esat-6* gene is not found in *M. bovis* BCG (Harboe *et al.*, 1996) since it maps in RD1, a 9.5-kb region absent from all *M. bovis* BCG strains (Mahairas *et al.*, 1996). Loss of RD1 is thought to be the first deletion event that occurred

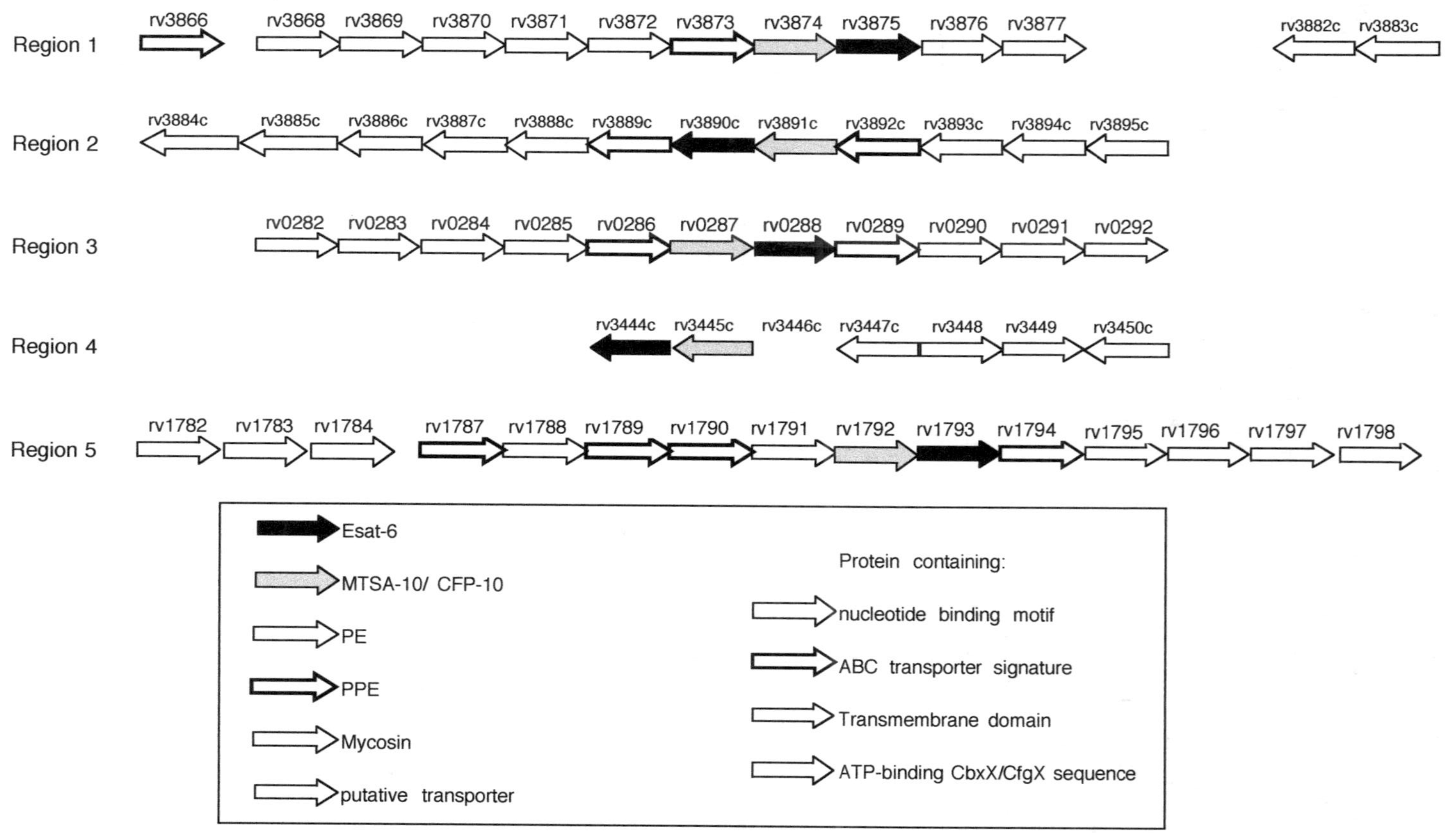

Figure 3. The five ESAT-6 gene clusters. Arrows having different colors/patterns represent different families of protein motifs as indicated. [Modified from (Gey van Pittius *et al.,* 2001)]

during serial passage of *M. bovis* by A. Calmette and C. Guerin (Behr *et al.*, 1999; Gordon *et al.*, 1999; Brosch *et al.*, 2000). Thus, deletion of RD1 might be associated with the loss of virulence in *M. bovis* BCG. However, complementation with RD1 of *M. bovis* BCG failed to restore virulence (Mahairas *et al.*, 1996).

Interestingly, the ESAT-6 antigen comes in pairs: in a search for the *esat-6* promoter, it was found that *esat-6* and the gene upstream (*lph*) were co-transcribed (Berthet *et al.*, 1998). The product of the *lhp* gene, called CFP-10 (Berthet *et al.*, 1998) or MTSA-10 (Colangeli *et al.*, 2000), was also found to elicit strong immune responses during TB in experimental animals (van Pinxteren *et al.*, 2000; Colangeli *et al.*, 2000) and in humans (Dillon *et al.*, 2000; Skjot *et al.*, 2000; Brusasca *et al.*, 2001; Doherty *et al.*, 2002). Thus, *M. tuberculosis* possesses two proteins, ESAT-6 and CFP-10/MTSA-10, that are encoded by adjacent, co-transcribed genes [Rv3874 and Rv3875 (Cole *et al.*, 1998)] and have similar molecular weights (~10 kDa). Both proteins are found in the culture filtrate (Sorensen *et al.*, 1995; Berthet *et al.*, 1998). Neither carries a cleavable, NH_2-terminal secretion signal peptide.

As the genome sequence of *M. tuberculosis* was analyzed it became evident that *esat-6* and *lhp* are present in multiple copies in the chromosome. Eleven duplications were described (Cole *et al.*, 1998). Six of them contain two to six genes, that is, the *esat-6/lhp* pair plus genes encoding members of the PE, PPE and QILSS protein families. These are antigens containing multiple copies of a Pro-Glu motif (PE) or a Pro-Pro-Glu motif (PPE) at their NH_2-termini, or a conserved QILSS sequence at their COOH termini (Cole *et al.*, 1998; Tekaia *et al.*, 1999). These ESAT-6 regions are usually flanked by one or two insertion sequences (Tekaia *et al.*, 1999), suggesting that they may have originated from a transposition event. The five larger duplications, referred to as the ESAT-6 loci (Tekaia *et al.*, 1999) or the ESAT-6 gene clusters (Gey Van Pittius *et al.*, 2001), include up to twelve genes (Figure 3). Their gene composition makes these clusters strongly reminiscent of functional units of the type I protein secretion system (Tekaia *et al.*, 1999; Gey Van Pittius *et al.*, 2001). Even though amino acid sequence homology analyses have not identified classical ABC transport systems in the ESAT-6 clusters (Braibant *et al.*, 2000), there is little doubt that these clusters may serve this function. As illustrated in Figure 3, four of five clusters include proteins bearing ABC transporter signatures. All of them also contain a large, integral-membrane protein bearing 11 transmembrane domains, and proteins having one or three nucleotide binding sites. Thus all the required components are present for a type I system dedicated to secretion of proteins of the ESAT-

6 family (N. Gey van Pittius, personal communication). In addition to the transport and energy-providing components of a type I secretion system, the five clusters include two other protein families (Tekaia *et al.*, 1999; Gey Van Pittius *et al.*, 2001). One is that of mycosins, cell-wall associated serine proteases (Brown *et al.*, 2000) (Figure 3) that may be involved in the processing of ESAT-6 family members (Gey Van Pittius *et al.*, 2001). The second family includes the ATP-binding CbxX/CfgX proteins (Figure 3), which tend to be located at either end of the cluster. A role as chaperones has been suggested for these proteins (Tekaia *et al.*, 1999).

The ESAT-6 gene clusters have been found in other mycobacteria and in other G+C-rich, Gram-positive bacteria (Gey Van Pittius *et al.*, 2001). Several findings were derived from such comparative genome analysis. First, copies of corresponding clusters show correspondingly similar flanking sequences, indicating that their genome location is conserved. Second, *M. leprae* contains functional copies of two clusters (regions 1 and 3 in Figure 3). The other regions are mostly present in *M. leprae* as pseudogenes, since they contain many non-sense or frame-shift mutations. Since the genome of *M. leprae* is thought to contain the minimal complement of genes defining pathogenic mycobacteria (Cole *et al.*, 2001), the conservation of two functional clusters could be taken to indicate a role for the ESAT-6 and CFP-10/MTSA10 antigens in survival of these pathogens. Third, of other G+C-rich, Gram-positive bacteria, *Corynebacterium diphteriae* contains a copy of region 4 only, and *Streptomyces coelicolor* carries ESAT-6 gene cluster orthologs that bear the most homology with region 4 genes. Thus, region 4 appears to be the ancestral region among the five ESAT-6 gene clusters of *M. tuberculosis* (Gey Van Pittius *et al.*, 2001). Finally, the presence of the clusters appears to be restricted to *Actinobacteria*, the class comprising all the G+C-rich bacteria mentioned above.

Further analysis of ESAT-6 and its homologs has recently led to the definition of the superfamily of WXG100 proteins, so called because they are proteins of ~100 amino acids in length and contain a central WXG motif (Pallen, 2002). The WXG100 protein family extends to Gram-positive bacteria outside *Actinobacteria*, where they appear to be involved in protein transport systems (Pallen, 2002).

In conclusion, the proteins of the ESAT-6 family are related in primary sequence, are usually found in pairs of coordinately expressed genes, and may serve a similar function. Additionally, they are usually found adjacent to proteins resembling the components of a dedicated type I protein secretion system. Given that the protein substrate is usually located close to the ABC transporter gene, it might be possible to identify all transporters associated with export of proteins in the ESAT-6 family.

Same Function, Different Gene Locations: the Antigen 85 Complex

It is fair to say that no protein of *M. tuberculosis* has been studied as extensively as the antigen 85 complex (Wiker *et al.*, 1992). The complex comprises three closely related enzymes named Antigen (Ag) 85A (31 kDa), Ag85B (30 kDa) and Ag85C (31.5 kDa) (Wiker *et al.*, 1986). Together, the three Ag85 proteins represent the predominant secreted protein species of tubercle bacilli cultivated *in vitro* [~25% of the culture filtrate (Wiker *et al.*, 1992; Harth *et al.*, 1996)] or found inside human mononuclear phagocytes

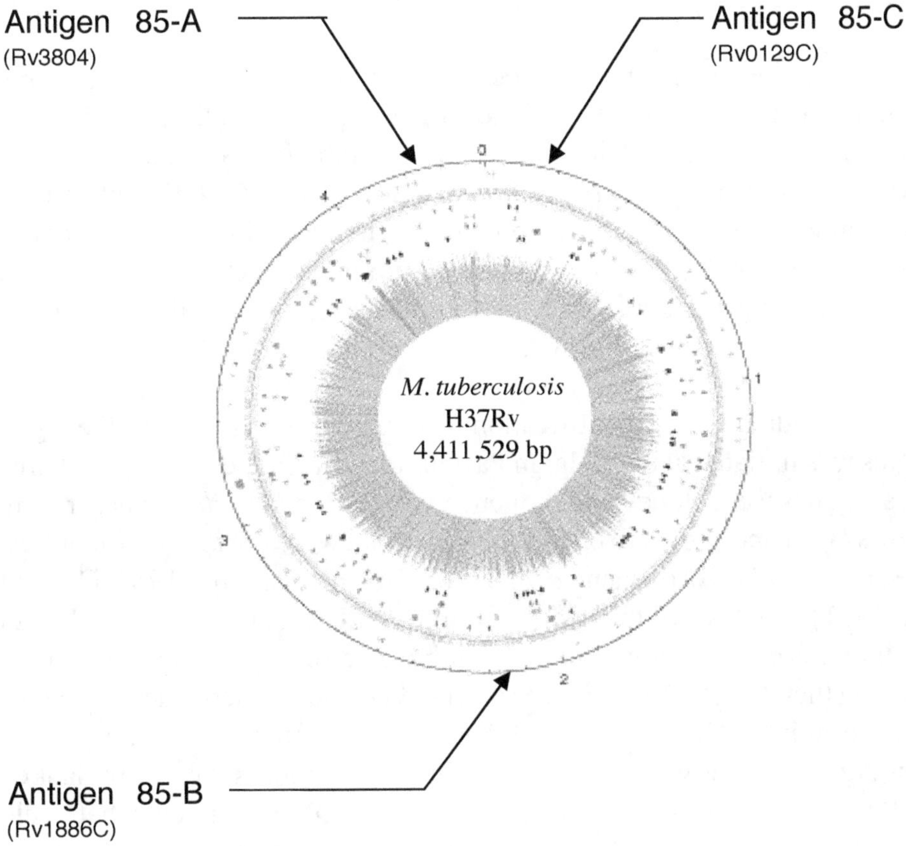

Figure 4. Location of the genes encoding Ag 85A, Ag 85B and Ag 85C in the genome of *M. tuberculosis* $H_{37}Rv$. The schematic representation of the genome was taken from the Tuberculist website (www.pasteur.fr/Bio/TubercuList).

(Lee *et al.*, 1995). Because of its abundance in the culture filtrate of *M. tuberculosis*, the Ag85 complex has been extensively evaluated as a vaccine candidate. Proteins of the Ag85 complex elicit protective immunity in animal models when administered either as a subunit vaccine (Horwitz *et al.*, 1995) as a DNA vaccine (Huygen *et al.*, 1996), or as a recombinant *M. bovis* BCG vaccine (Horwitz *et al.*, 2000).

The function of the Ag85 proteins has also been extensively studied (Daffe, 2000). Each protein of the complex possesses the mycolyltransferase activity (Belisle *et al.*, 1997) that is required for synthesis of α-α'-trehalose dimycolate (cord factor), a key component of the cell wall of *M. tuberculosis*. The role of the Ag85 enzymes in an important biosynthetic pathway, such as that of cell wall biogenesis, makes them attractive targets for drug design. In this light, further investigation on the function of these enzymes is ongoing (Daffe, 2000) and the crystal structures of Ag85C and Ag85 B have been solved (Sharma *et al.*, 2000; Anderson *et al.*, 2001).

The reason why *M. tuberculosis* needs three enzymes to serve the same function remains unclear. Inactivation of the gene encoding Ag85A was associated with poor growth in minimal, semi-synthetic media and in cultured macrophages (Armitige *et al.*, 2000), while an Ag85C-deficient mutant showed alterations in mycolate content and permeability of the cell wall (Jackson *et al.*, 1999). However, these single gene inactivations were not lethal, suggesting perhaps that it may be necessary to eliminate all three Ag85 proteins to arrest cell growth.

The work leading to the identification of three distinct genes for the Ag85 proteins was initiated at the Belgian Pasteur Institute in the late 1980s. Three separate genes were cloned and sequenced from *M. tuberculosis* and/or from *M. bovis* BCG and shown to encode proteins that were 68% to 80% identical at the amino acid level (Matsuo *et al.*, 1988; Borremans *et al.*, 1989; Content *et al.*, 1991). Analysis of the deduced amino acid sequences showed that all three proteins carried a long (40 to 46 amino acid residues) secretion signal, indicating that the Ag85 proteins are secreted via the *sec*-dependent pathway. The same authors (Kremer *et al.*, 1995) also showed that the structural genes for the Ag85 proteins were not clustered in the genome, as they were flanked by different nucleotide sequences and mapped in DNA fragments that could be separated by pulse-field gel electrophoresis. Similar conclusions were reached also by another group (Harth *et al.*, 1996). Indeed, the Ag85 proteins are encoded by genes located distantly on the *M. tuberculosis* genome (Rv3804c for Ag85A, Rv1886c for Ag85B and Rv0129c for Ag85C) (Figure

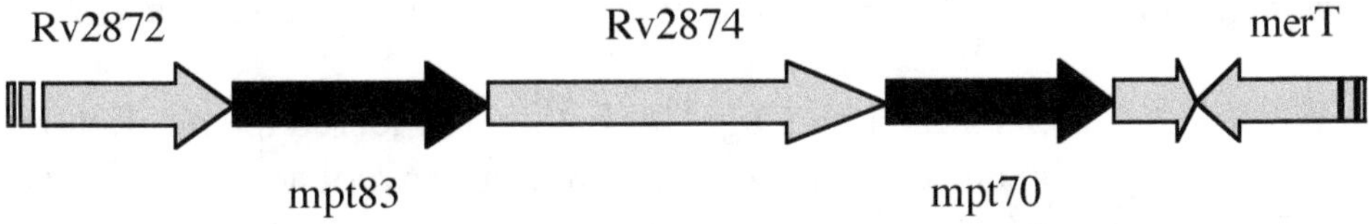

Lipoprotein signal peptide cleavage site

```
MPT83:       MINVQAKPAAAASLAAIAIAFLAGCSSTKPVSQDTSPKPATSPAAPVTTAAMADPAA
                                          +   +   A   AA +  +      A+
MPT70:                                    MKVKNTIAATSFAAAGLAALAVAVSPPAAAG

MPT83: 58   DLIGRGCAQYAAQNPTGPGSVAGMAQDPVATAASNNPMLSTLTSALSGKLNPDVNLVDTL
            DL+G GCA+YAA NPTGP SV GM+QDPVA AASNNP L+TLT+ALSG+LNP VNLVDTL
MPT70: 32   DLVGPGCAEYAAANPTGPASVQGMSQDPVAVAASNNPELTTLTAALSGQLNPQVNLVDTL

MPT83: 118  NGGEYTVFAPTNAAFDKLPAATIDQLKTDAKLLSSILTYHVIAGQASPSRIDGTHQTLQG
            N G+YTVFAPTNAAF KLPA+TID+LKT++ LL+SILTYHV+AGQ SP+ + GT QTLQG
MPT70: 92   NSGQYTVFAPTNAAFSKLPASTIDELKTNSSLLTSILTYHVVAGQTSPANVVGTRQTLQG

MPT83: 178  ADLTVIGARDDLMVNNAGLVCGGVHTANATVYMIDTVLMPPA 219
            A +TV G   + L V NA +VCGGV TANATVYMID+VLMPPA
MPT70: 152  ASVTVTGQGNSLKVGNADVVCGGVSTANATVYMIDSVLMPPA 193
```

Figure 5. Location of the genes encoding MPT70 and MPT83 in the genome of *M. tuberculosis* $H_{37}Rv$ and alignment of the relative amino acid sequences.

4). It has also been reported that the three Ag85 proteins were present in the culture supernatant at a ratio of 3:2:1 (Ag85B:Ag85A:Ag85C) over several weeks of growth *in vitro* (Harth *et al.*, 1996), suggesting that the three corresponding genes may be coordinately regulated. At present, however, there is no definitive evidence supporting this possibility as analysis of upstream sequences has failed to identify common regulatory elements for the three genes encoding the Ag85 proteins (Kremer *et al.*, 1995).

In conclusion, the proteins of the Ag85 family are very closely related in amino acid sequence and in structure and serve a similar function. However, they are not encoded by clustered genes.

Homologous Proteins, Different Subcellular Locations: MPB70 and MPB83

Credit must be given to Sadamu Nagai for having pioneered the purification of proteins from the culture filtrate of tuberculous mycobacteria. One was MPB70 (Nagai *et al.*, 1981), a secreted protein of ~ 22 kDa in molecular weight that was found to constitute 10% of the culture filtrate proteins of *M. bovis* BCG Tokyo (Harboe *et al.*, 1984). It was soon realized, however, that MPB70 was produced in widely varying amounts by different substrains of *M. bovis* BCG, with substrains Tokyo and Sweden representing some of the high producers and substrains Pasteur and Copenhagen representing some of the low producers (Miura *et al.*, 1983; Harboe *et al.*, 1984; Matsuo *et al.*, 1995). Subsequent work (Harboe *et al.*, 1995) led to the identification of another protein in the culture fluid of *M. bovis* BCG Tokyo. This second protein, named MPB83, was highly homologous to, but distinct from, MPB70 (Harboe *et al.*, 1995) (Figure 5). Interestingly, the pattern of low and high producing strains described for MPB70 was the same for MPB83 (Hewinson *et al.*, 1996; Wiker *et al.*, 1996). However, while low MPB70 producers also produced little MPB83, high MPB70 producers were relatively heterogeneous in their level of MPB83 synthesis (Wiker *et al.*, 1996). With the use of gene cloning and sequencing it was confirmed that the genes for MPB70 and for MPB83 were distinct, mapped 2.4 kb apart, and were transcribed in the same direction (Terasaka *et al.*, 1989; Matsuo *et al.*, 1996). The same gene sequences and organization were later found in *M. tuberculosis* (in this organism, the two proteins are called MPT70 and MPT83) (Matsumoto *et al.*, 1995; Hewinson *et al.*, 1996). However, *M. tuberculosis* cultures produce little, if any, of either protein (Harboe *et al.*, 1984; Hewinson *et al.*, 1996; Matsuo *et al.*, 1996). Nucleotide sequence analyses of the *mpb/t70* and *mpb/t83* genes indicated that the two gene products are highly homologous, since (i) both genes encode cleavable, NH_2-terminal signal peptides and (ii) the deduced amino acid sequences of the two mature proteins are 63% identical (Figure 5). However, two major differences were observed. One is that the *mpb/t83* gene encodes a lipoprotein signal peptide cleavage sequence while the *mpb/t70* gene does not (Hewinson *et al.*, 1996; Matsuo *et al.*, 1996; Vosloo *et al.,* 1997). The second difference lies in the presence in *mpb/t83* of an extra sequence encoding an NH_2-terminal, 35 amino acid-long glycosylation site within the mature protein (Hewinson *et al.*, 1996). Thus, MPB/T83 should be a glycolipoprotein exposed on the mycobacterial cell surface, while MPB/T70 should be an exoprotein. The availability of monoclonal antibodies specific to either protein (Harboe *et al.*, 1998; Wiker *et al.*, 1998; and references therein) made it possible to test the subcellular localization of MPB70 and MPB83. In accordance with the sequence data,

MPB70 was found to be predominantly extracellular (Wiker *et al.*, 1991). MPB83 was shown to exist in two forms migrating with different apparent molecular masses, a 26 kDa species that remained cell-associated and a 23 kDa form that was mostly found in the culture fluid (Harboe *et al.*, 1997; Harboe *et al.*, 1998). The differences between the two forms of the MPB83 are that the 23 kDa species is not lipidated (Harboe *et al.*, 1998; Harboe *et al.,* 2002) and that the signal sequence is processed at slightly different sites (M. Harboe, personal communication).

Consistent with their subcellular localization (extracellular and surface-associated), MPB70 and MPB83 have been shown to induce host immune responses of value for TB immunodiagnostics and vaccine development [examples are (Nagai *et al.*, 1981; Haslov *et al.*, 1987; Billman-Jacobe *et al.*, 1990; Harboe *et al.*, 1990; Pollock *et al.*, 1994; Roche *et al.*, 1994; Vordermeier *et al.*, 1999; Pollock *et al.*, 2000; Vordermeier *et al.*, 2000; Chambers *et al.*, 2001)]. It is worth mentioning that their *M. tuberculosis* counterparts are also highly immunogenic during experimental infection with live bacilli [(Hewinson *et al.*, 1996); K. Lyashchenko and M.L. Gennaro, unpublished observations], even though the corresponding genes are expressed at low levels in *M. tuberculosis* cultures *in vitro*. The function of these antigens remains unknown, but a similarity with the osteoblast-specific factor 2 (OSF-2) (Matsumoto *et al.*, 1995) has been taken to suggest that the two proteins could be involved in the bone tropism of mycobacteria (Ulstrup *et al.*, 1995).

That the two proteins are both produced at a similar level (low or high) by different strains of tubercle bacilli is suggestive of co-regulation of their corresponding genes. A support of this possibility derives from the observation that *mpt70* was expressed in *Mycobacterium smegmatis* only when cloned as a DNA fragment containing both *mpt70* and *mpt83* genes (Hewinson *et al.*, 1996). A recent study (Juarez *et al.*, 2001) suggests that the two genes, and the intervening DNA, are transcribed by promoter sequences located upstream of *mpt83*.

In conclusion, the MPB70 and MPB83 proteins are highly related in primary sequence and are encoded by closely located, presumably co-regulated genes. However, they localize to different subcellular compartments. Whether their different localization in tubercle bacilli may reflect functional differences remains unknown.

Concluding Remarks

Bacteria have evolved at least four pathways of protein secretion. Only one, the type II system, is obviously marked in the form of a cleavable, NH_2-terminal signal peptide. It is still unknown whether genomic analyses can overcome the lack of an obvious marker for protein secretion by the other pathways. The examples we have presented of three families of secreted proteins of *M. tuberculosis* clearly show that there are no simple rules that can predict the fate of a protein in terms of its secretion. Two secreted proteins, ESAT-6 and CFP10/MTSA10, may be encoded by genes that are adjacent and co-expressed, and probably also share the same dedicated secretion system. But two other proteins, MPB83 and MPB70, may be encoded by genes that are highly homologous, closely located, and coordinately regulated, and yet be destined to different subcellular compartments and presumably to different functions. Still other secreted proteins, members of Ag85 complex, may have essentially the same function and be encoded by genes located far from each other. Despite the present uncertainties, however, the possibility of expanding the analysis of homologous protein sequences and conserved motifs to the ever increasing number of sequenced bacterial genomes will help define novel gene clusters representing type I, type III and type IV secretion systems. Moreover, as with the ESAT-6 clusters, finding the substrates opens the way to finding the transporters. Conversely, identifying transporters will help discover their substrates. And with an increasing number of identified secreted proteins, comparative genomic analyses will help define the nature of those yet elusive markers of type I, type II and type IV secretion. It is only a matter of time.

Acknowledgements

We wish to thank N. Gey van Pittius and M. Pallen for sharing information prior to publication; N. Gey van Pittius, M. Harboe and K. Drlica for critical reading of the manuscript or parts thereof. This work was supported by NIH grant AI-36989. We wish to dedicate this chapter to the children of New York City who lost their parents on September 11, 2001.

References

Aldridge, P. and Hughes, K.T. 2001. How and when are substrates selected for type III secretion? Trends Microbiol. 9: 209-214.

Amara, R.R. and Satchidanandam, V. 1996. Analysis of a genomic DNA expression library of *Mycobacterium tuberculosis* using tuberculosis patient sera: evidence for modulation of host immune response. Infect. Immun. 64: 3765-3771.

Andersen, A.B. and Brennan, P.J. 1994. Proteins and antigens of *Mycobacterium tuberculosis*. In: Tuberculosis: pathogenesis, protection, and control. B.R. Bloom, ed. ASM Press, Washington, DC. p.307-327.

Andersen, P. 1994. The T cell response to secreted antigens of *Mycobacterium tuberculosis*. Immunobiology. 191: 537-547.

Andersen, P. 1997. Host responses and antigens involved in protective immunity to *Mycobacterium tuberculosis*. Scand. J. Immunol. 45: 115-131.

Andersen, P., Askgaard, D., Gottschau, A., Bennedsen, J., Nagai, S. and Heron, I. 1992. Identification of immunodominant antigens during infection with *Mycobacterium tuberculosis*. Scand. J. Immunol. 36: 823-831.

Andersen, P., Askgaard, D., Ljungqvist, L., Bennedsen, J. and Heron, I. 1991. Proteins released from *Mycobacterium tuberculosis* during growth. Infect. Immun. 59: 1905-1910.

Anderson, D.H., Harth, G., Horwitz, M.A. and Eisenberg, D. 2001. An interfacial mechanism and a class of inhibitors inferred from two crystal structures of the *Mycobacterium tuberculosis* 30 kDa major secretory protein (Antigen 85B), a mycolyl transferase. J. Mol. Biol. 307: 671-681.

Anderson, D.M. and Schneewind, O. 1997. A mRNA signal for the type III secretion of Yop proteins by *Yersinia enterocolitica*. Science. 278: 1140-1143.

Anderson, D.M. and Schneewind, O. 1999. *Yersinia enterocolitica* type III secretion: an mRNA signal that couples translation and secretion of YopQ. Mol. Microbiol. 31: 1139-1148.

Arend, S.M., Geluk, A., van Meijgaarden, K.E., van Dissel, J.T., Theisen, M., Andersen, P. and Ottenhoff, T.H. 2000. Antigenic equivalence of human T-cell responses to *Mycobacterium tuberculosis*-specific RD1-encoded protein antigens ESAT-6 and culture filtrate protein 10 and to mixtures of synthetic peptides. Infect. Immun. 68: 3314-3321.

Armitige, L.Y., Jagannath, C., Wanger, A.R. and Norris, S.J. 2000. Disruption of the genes encoding antigen 85A and antigen 85B of *Mycobacterium tuberculosis* H37Rv: effect on growth in culture and in macrophages. Infect. Immun. 68: 767-778.

Behr, M.A., Wilson, M.A., Gill, W.P., Salamon, H., Schoolnik, G.K., Rane, S. and Small, P.M. 1999. Comparative genomics of BCG vaccines by whole-genome DNA microarray. Science. 284: 1520-1523.

Belisle, J.T., Vissa, V.D., Sievert, T., Takayama, K., Brennan, P.J. and Besra, G.S. 1997. Role of the major antigen of *Mycobacterium tuberculosis* in cell wall biogenesis. Science. 276: 1420-1422.

Bennett, J.C. and Hughes, C. 2000. From flagellum assembly to virulence: the extended family of type III export chaperones. Trends Microbiol. 8: 202-204.

Berthet, F.X., Rasmussen, P.B., Rosenkrands, I., Andersen, P. and Gicquel, B. 1998. A *Mycobacterium tuberculosis* operon encoding ESAT-6 and a novel low-molecular-mass culture filtrate protein (CFP-10). Microbiology. 144: 3195-3203.

Bieker, K.L. and Silhavy, T.J. 1990. The genetics of protein secretion in *E. coli*. Trends Genet. 6: 329-334.

Billman-Jacobe, H., Radford, A.J., Rothel, J.S. and Wood, P.R. 1990. Mapping of the T and B cell epitopes of the *Mycobacterium bovis* protein, MPB70. Immunol. Cell Biol. 68: 359-365

Binet, R., Letoffe, S., Ghigo, J.M., Delepelaire, P. and Wandersman, C. 1997. Protein secretion by Gram-negative bacterial ABC exporters—a review. Gene. 192: 7-11.

Blight, M.A. and Holland, I.B. 1994. Heterologous protein secretion and the versatile *Escherichia coli* haemolysin translocator. Trends Biotechnol. 12: 450-455.

Bloch, H. and Segal, W. 1955. Viability and multiplication of vaccines in immunization against tuberculosis. Am. Rev. Tubercul. Pulm. Dis. 71: 228-245

Borremans, M., de Wit, L., Volckaert, G., Ooms, J., de Bruyn, J., Huygen, K., van Vooren, J.P., Stelandre, M., Verhofstadt, R. and Content, J. 1989. Cloning, sequence determination, and expression of a 32-kilodalton-protein gene of *Mycobacterium tuberculosis*. Infect. Immun. 57: 3123-3130.

Braibant, M., Gilot, P. and Content, J. 2000. The ATP binding cassette (ABC) transport systems of *Mycobacterium tuberculosis*. FEMS Microbiol. Rev. 24: 449-467.

Braunstein, M., Brown, A.M., Kurtz, S. and Jacobs, W.R., Jr. 2001. Two nonredundant SecA homologues function in mycobacteria. J. Bacteriol. 183: 6979-6990.

Brennan, P.J. and Draper, P. 1994. Ultrastructure of *Mycobacterium tuberculosis*. In: Tuberculosis: pathogenesis, protection, and control. B.R. Bloom, ed. ASM Press, Washington, DC. p. 271-284

Brosch, R., Gordon, S.V., Buchrieser, C., Pym, A.S., Garnier, T. and Cole, S.T. 2000. Comparative genomics uncovers large tandem chromosomal duplications in *Mycobacterium bovis* BCG Pasteur. Yeast. 17: 111-123.

Brown, G.D., Dave, J.A., Gey van Pittius, N.C., Stevens, L., Ehlers, M.R. and Beyers, A.D. 2000. The mycosins of *Mycobacterium tuberculosis* H37Rv: a family of subtilisin-like serine proteases. Gene. 254: 147-155.

Brusasca, P.N., Colangeli, R., Lyashchenko, K.P., Zhao, X., Vogelstein, M., Spencer, J.S., McMurray, D.N. and Gennaro, M.L. 2001. Immunological characterization of antigens encoded by the RD1 region of the *Mycobacterium tuberculosis* genome. Scand. J. Immunol. 54: 448-452.

Buchanan, S.K. 2001. Type I secretion and multidrug efflux: transport through the TolC channel-tunnel. Trends Biochem. Sci. 26: 3-6.

Burns, D.L. 1999. Biochemistry of type IV secretion. Curr. Opin. Microbiol. 2: 25-29.

Chambers, M.A., Stagg, D., Gavier-Widen, D., Lowrie, D., Newell, D. and Hewinson, R.G. 2001. A DNA encoding MPB83 from *Mycobacterium bovis* reduces *M. bovis* dissemination to the kidneys of mice and is expressed in primary cell cultures of the European badger (Meles meles). Res. Vet. Sci. 71: 119-126

Cheng, L.W. and Schneewind, O. 2000. Type III machines of Gram-negative bacteria: delivering the goods. Trends Microbiol. 8: 214-220.

Christie, P.J. and Vogel, J.P. 2000. Bacterial type IV secretion: conjugation systems adapted to deliver effector molecules to host cells. Trends Microbiol. 8: 354-360.

Christie, P.J. 2001. Type IV secretion: intercellular transfer of macromolecules by systems ancestrally related to conjugation machines. Mol. Microbiol. 40: 294-305.

Chubb, A.J., Woodman, Z.L., da Silva Tatley, F.M., Hoffmann, H.J., Scholle, R.R. and Ehlers, M.R. 1998. Identification of *Mycobacterium tuberculosis* signal sequences that direct the export of a leaderless beta-lactamase gene product in *Escherichia coli*. Microbiology. 144: 1619-1629.

Colangeli, R., Spencer, J.S., Bifani, P., Williams, A., Lyashchenko, K., Keen, M.A., Hill, P.J., Belisle, J. and Gennaro, M.L. 2000. MTSA-10, the product of the Rv3874 gene of *Mycobacterium tuberculosis*, elicits tuberculosis-specific, delayed-type hypersensitivity in guinea pigs. Infect. Immun. 68: 990-993.

Cole, S.T., Brosch, R., Parkhill, J., Garnier, T., Churcher, C., Harris, D., Gordon, S.V., Eiglmeier, K., Gas, S., Barry, C.E., 3rd, Tekaia, F., Badcock, K., Basham, D., Brown, D., Chillingworth, T., Connor, R., Davies, R., Devlin, K., Feltwell, T., Gentles, S., Hamlin, N., Holroyd, S., Hornsby, T., Jagels, K., Barrell, B.G. and et al. 1998. Deciphering the biology of *Mycobacterium tuberculosis* from the complete genome sequence. Nature. 393: 537-544.

Cole, S.T., Eiglmeier, K., Parkhill, J., James, K.D., Thomson, N.R., Wheeler, P.R., Honore, N., Garnier, T., Churcher, C., Harris, D., Mungall, K., Basham, D., Brown, D., Chillingworth, T., Connor, R., Davies, R.M., Devlin, K., Duthoy, S., Feltwell, T., Fraser, A., Hamlin, N., Holroyd, S., Hornsby, T.,

Jagels, K., Lacroix, C., Maclean, J., Moule, S., Murphy, L., Oliver, K., Quail, M.A., Rajandream, M.A., Rutherford, K.M., Rutter, S., Seeger, K., Simon, S., Simmonds, M., Skelton, J., Squares, R., Squares, S., Stevens, K., Taylor, K., Whitehead, S., Woodward, J.R., Barrell, B.G., Demissie, A., Ravn, P., Olobo, J., Doherty, T.M., Eguale, T., Geletu, M., Hailu, W., Andersen, P. and Britton, S. 2001. Massive gene decay in the leprosy bacillus. Nature. 409: 1007-1011.

Collins, H.L. and Kaufmann, S.H. 2001. The many faces of host responses to tuberculosis. Immunology. 103: 1-9.

Content, J., de la Cuvellerie, A., De Wit, L., Vincent-Levy-Frebault, V., Ooms, J. and De Bruyn, J. 1991. The genes coding for the antigen 85 complexes of *Mycobacterium tuberculosis* and *Mycobacterium bovis* BCG are members of a gene family: cloning, sequence determination, and genomic organization of the gene coding for antigen 85-C of M. tuberculosis. Infect. Immun. 59: 3205-3212.

Cooper, A.M. and Flynn, J.L. 1995. The protective immune response to *Mycobacterium tuberculosis*. Curr. Opin. Immunol. 7: 512-516.

Cornelis, G.R. and Van Gijsegem, F. 2000. Assembly and function of type III secretory systems. Ann. Rev. Microbiol. 54: 735-774.

Daffe, M. and Etienne G. 1999. The capsule of *Mycobacterium tuberculosis* and its implications for pathogenicity. Tuber. Lung. Dis. 79: 153-169.

Daffe, M. 2000. The mycobacterial antigens 85 complex - from structure to function and beyond. Trends Microbiol. 8: 438-440.

Danese, P.N. and Silhavy, T.J. 1998. Targeting and assembly of periplasmic and outer-membrane proteins in *Escherichia coli*. Ann. Rev. Genet. 32: 59-94.

Delepelaire, P. and Wandersman, C. 1990. Protein secretion in gram-negative bacteria. The extracellular metalloprotease B from *Erwinia chrysanthemi* contains a C-terminal secretion signal analogous to that of *Escherichia coli* alpha-hemolysin. J. Biol. Chem. 265: 17118-17125.

Dillon, D.C., Alderson, M.R., Day, C.H., Bement, T., Campos-Neto, A., Skeiky, Y.A., Vedvick, T., Badaro, R., Reed, S.G. and Houghton, R. 2000. Molecular and immunological characterization of *Mycobacterium tuberculosis* CFP-10, an immunodiagnostic antigen missing in *Mycobacterium bovis* BCG. J. Clin. Microbiol. 38: 3285-3290.

Dinh, T., Paulsen, I.T. and Saier, M.H., Jr. 1994. A family of extracytoplasmic proteins that allow transport of large molecules across the outer membranes of gram-negative bacteria. J. Bacteriol. 176: 3825-3831.

Doherty, T.M., Demissie, A., Olobo, J., Wolday, D., Britton, S., Eguale, T., Ravn, P. and Andersen, P. 2002. Immune responses to the *Mycobacterium tuberculosis*-specific antigen ESAT-6 signal subclinical infection among contacts of tuberculosis patients. J. Clin. Microbiol. 40: 704-706.

Economou, A. 1998. Bacterial preprotein translocase: mechanism and conformational dynamics of a processive enzyme. Mol. Microbiol. 27: 511-518.

Fath, M.J., Skvirsky, R.C. and Kolter, R. 1991. Functional complementation between bacterial MDR-like export systems: colicin V, alpha-hemolysin, and Erwinia protease. J. Bacteriol. 173: 7549-7556.

Felmlee, T., Pellett, S., and Welch, R.A. 1985. Nucleotide sequence of an *Escherichia coli* chromosomal hemolysin. J. Bacteriol. 163: 94-105.

Finlay, B.B. and Falkow, S. 1997. Common themes in microbial pathogenicity revisited. Microbiol. Mol. Biol. Rev. 61: 136-169.

Genin, S. and Boucher, C.A. 1994. A superfamily of proteins involved in different secretion pathways in gram-negative bacteria: modular structure and specificity of the N-terminal domain. Mol Gen Genet. 243: 112-118.

Gentschev, I., Dietrich, G. and Goebel, W. 2002. The *Escherishia coli* α-hemolysin secretion system and its use in vaccine development. Trends Microbiol. 10: 39-45.

Gey Van Pittius, N.C., Gamieldien, J., Hide, W., Brown, G.D., Siezen, R.J. and Beyers, A.D. 2001. The ESAT-6 gene cluster of *Mycobacterium tuberculosis* and other high G+C Gram-positive bacteria. Genome Biol. 2: RESEARCH0044.

Gilmore, M.S., Segarra, R.A. and Booth, M.C. 1990. An HlyB-type function is required for expression of the *Enterococcus faecalis* hemolysin/bacteriocin. Infect. Immun. 58: 3914-3923.

Gomez, M., Johnson, S. and Gennaro, M.L. 2000. Identification of secreted proteins of *Mycobacterium tuberculosis* by a bioinformatic approach. Infect. Immun. 68: 2323-2327.

Gordon, S.V., Brosch, R., Billault, A., Garnier, T., Eiglmeier, K. and Cole, S.T. 1999. Identification of variable regions in the genomes of tubercle bacilli using bacterial artificial chromosome arrays. Mol. Microbiol. 32: 643-655.

Groisman, E.A. and Ochman, H. 1997. How Salmonella became a pathogen. Trends Microbiol. 5: 343-349.

Harboe, M. and Nagai, S. 1984. MPB70, a unique antigen of *Mycobacterium bovis* BCG. Am. Rev. Tubercul. Pulm. Dis. 129: 444-452

Harboe, M., Nagai, S., Wiker, H.G., Sletten, K. and Haga, S. 1995. Homology between the MPB70 and MPB83 proteins of *Mycobacterium bovis* BCG. Scand J. Immunol. 42: 46-51.

Harboe, M., Oettinger, T., Wiker, H.G., Rosenkrands, I. and Andersen, P. 1996. Evidence for occurrence of the ESAT-6 protein in *Mycobacterium tuberculosis* and virulent *Mycobacterium bovis* and for its absence in

Mycobacterium bovis BCG. Infect. Immun. 64: 16-22.

Harboe, M. and Wiker, H.G. 1997. Different localization of the MPB70 and the MPB80 proteins of *Mycobacterium bovis*. Med. Princ. Pract. 6: 84-90

Harboe, M., Wiker, H.G., Duncan, J.R., Garcia, T.W., Dukes, T.W., Brooks, C., Turcotte, C. and Nagai, S. 1990. Protein G-based enzyme-linked immunosorbent assay for anti-MPB70 antibodies in bovine tuberculosis. J. Clin. Microbiol. 28: 913-921

Harboe, M., Wiker, H.G., Ulvund, G., Lund-Pedersen, B., Andersen, A.B., Hewinson, R.G. and Nagai, S. 1998. MPB70 and MPB83 as indicators of protein localization in mycobacterial cells. Infect. Immun. 66: 289-296

Harboe, M., Whelan, A.O., Ulvund, G., McNair, J., Pollock J.M., Hewinson R.G. and Wiker, H.G. 2002. Generation of antibodies to the signal peptide of the MPT83 lipoprotein of *Mycobacterium tuberculosis*. Scand. J. Immunol. 55: 82-87

Harth, G., Lee, B.Y., Wang, J., Clemens, D.L. and Horwitz, M.A. 1996. Novel insights into the genetics, biochemistry, and immunocytochemistry of the 30-kilodalton major extracellular protein of *Mycobacterium tuberculosis*. Infect. Immun. 64: 3038-3047.

Haslov, K., Andersen, A.B. and M.W., B. 1987. Biological activity in sensitized guinea pigs of MPB70, a protein specific forsome strains of *Mycobacterium bovis* BCG. Scand J. Immunol. 26: 445-454

Hewinson, R.G., Michell, S.L., Russell, W.P., McAdam, R.A. and Jacobs, W.R., Jr. 1996. Molecular characterization of MPT83: a seroreactive antigen of *Mycobacterium tuberculosis* with homology to MPT70. Scand J. Immunol. 43: 490-499.

Higgins, C.F. 1992. ABC transporters: from microorganisms to man. Ann. Rev. Cell Biol. 8: 67-113.

Hirst, T.R., Sanchez, J., Kaper, J.B., Hardy, S.J. and Holmgren, J. 1984. Mechanisms of toxin secretion by Vibrio cholerae investigated in strains harboring plasmids that encode heat-labile enterotoxins of *Escherichia coli*. Proc. Natl. Acad. Sci. USA. 81: 7752-7756.

Horwitz, M.A., Harth, G., Dillon, B.J. and Maslesa-Galic, S. 2000. Recombinant bacillus calmette-guerin (BCG) vaccines expressing the *Mycobacterium tuberculosis* 30-kDa major secretory protein induce greater protective immunity against tuberculosis than conventional BCG vaccines in a highly susceptible animal model. Proc. Natl. Acad. Sci. USA. 97: 13853-13858.

Horwitz, M.A., Lee, B.W., Dillon, B.J. and Harth, G. 1995. Protective immunity against tuberculosis induced by vaccination with major extracellular proteins of *Mycobacterium tuberculosis*. Proc. Natl. Acad. Sci. USA. 92: 1530-1534.

Hueck, C.J. 1998. Type III protein secretion systems in bacterial pathogens of animals and plants. Microbiol Mol Biol Rev. 62: 379-433.

Huygen, K., Content, J., Denis, O., Montgomery, D.L., Yawman, A.M., Deck, R.R., DeWitt, C.M., Orme, I.M., Baldwin, S., D'Souza, C., Drowart, A., Lozes, E., Vandenbussche, P., Van Vooren, J.P., Liu, M.A. and Ulmer, J.B. 1996. Immunogenicity and protective efficacy of a tuberculosis DNA vaccine. Nat. Med. 2: 893-898.

Jackson, M., Raynaud, C., Laneelle, M.A., Guilhot, C., Laurent-Winter, C., Ensergueix, D., Gicquel, B. and Daffe, M. 1999. Inactivation of the antigen 85C gene profoundly affects the mycolate content and alters the permeability of the *Mycobacterium tuberculosis* cell envelope. Mol. Microbiol. 31: 1573-1587.

Juarez, M.D., Torres, A. and Espitia, C. 2001. Characterization of the *Mycobacterium tuberculosis* region containing the mpt83 and mpt70 genes. FEMS Microbiol Lett. 203: 95-102

Jungblut, P.R. 2001. Proteome analysis of bacterial pathogens. Microbes Infect. 3: 831-840.

Kamath, A.T., Feng, C.G., MacDonald, M., Briscoe, H., Britton, W.J.1999. Differential protective efficacy of DNA vaccines expressing secreted proteins of *Mycobacterium tuberculosis*. Infect. Immun. 67: 1702-1707

Kremer, L., Baulard, A., Estaquier, J., Content, J., Capron, A. and Locht, C. 1995. Analysis of the *Mycobacterium tuberculosis* 85A antigen promoter region. J. Bacteriol. 177: 642-653.

Kubori, T., Matsushima, Y., Nakamura, D., Uralil, J., Lara-Tejero, M., Sukhan, A., Galan, J.E. and Aizawa, S.I. 1998. Supramolecular structure of the *Salmonella typhimurium* type III protein secretion system. Science. 280: 602-605.

Lee, B.Y. and Horwitz, M.A. 1995. Identification of macrophage and stress-induced proteins of *Mycobacterium tuberculosis*. J. Clin Invest. 96: 245-249.

Letoffe, S., Delepelaire, P. and Wandersman, C. 1990. Protease secretion by *Erwinia chrysanthemi*: the specific secretion functions are analogous to those of *Escherichia coli* alpha-haemolysin. EMBO J. 9: 1375-1382.

Letoffe, S. and Wandersman, C. 1992. Secretion of CyaA-PrtB and HlyA-PrtB fusion proteins in *Escherichia coli*: involvement of the glycine-rich repeat domain of *Erwinia chrysanthemi* protease B. J. Bacteriol. 174: 4920-4927.

Lim, E.M., Rauzier, J., Timm, J., Torrea, G., Murray, A., Gicquel, B. and Portnoi, D. 1995. Identification of *Mycobacterium tuberculosis* DNA sequences encoding exported proteins by using *phoA* gene fusions. J. Bacteriol. 177: 59-65.

Limia, A., Sangari, F.J., Wagner, D. and Bermudez, L.E. 2001. Characterization and expression of *secA* in *Mycobacterium avium*. FEMS Microbiol Lett. 197: 151-157.

Linton, K.J., and Higgins, C.F. 1998. The *Escherichia coli* ATP-binding cassette (ABC) proteins. Mol. Microbiol. 28: 5-13.

Liu, J., Barry, C.E., 3rd, Besra, G.S. and Nikaido, H. 1996. Mycolic acid structure determines the fluidity of the mycobacterial cell wall. J. Biol. Chem. 271: 29545-51.

Lu, H.M. and Lory, S. 1996. A specific targeting domain in mature exotoxin A is required for its extracellular secretion from *Pseudomonas aeruginas*. EMBO J. 15: 429-436

Lyashchenko, K., Colangeli, R., Houde, M., Al Jahdali, H., Menzies, D. and Gennaro, M.L. 1998. Heterogeneous antibody responses in tuberculosis. Infect. Immun. 66: 3936-3940.

Mackman, N., Nicaud, J.M., Gray, L. and Holland, I.B. 1986. Secretion of haemolysin by *Escherichia coli*. Curr. Top. Microbiol. Immunol. 125: 159-181.

Mahairas, G.G., Sabo, P.J., Hickey, M.J., Singh, D.C. and Stover, C.K. 1996. Molecular analysis of genetic differences between *Mycobacterium bovis* BCG and virulent *M. bovis*. J. Bacteriol. 178: 1274-1282.

Manca, C., Lyashchenko, K., Wiker, H.G., Usai, D., Colangeli, R. and Gennaro, M.L. 1997. Molecular cloning, purification, and serological characterization of MPT63, a novel antigen secreted by *Mycobacterium tuberculosis*. Infect. Immun. 65: 16-23.

Matsumoto, S., Matsuo, T., Ohara, N., Hotokezaka, H., Naito, M., Minami, J. and Yamada, T. 1995. Cloning and sequencing of a unique antigen MPT70 from *Mycobacterium tuberculosis* H37Rv and expression in BCG using *E. coli*-mycobacteria shuttle vector. Scand. J. Immunol. 41: 281-287.

Matsuo, K., Yamaguchi, R., Yamazaki, A., Tasaka, H., Yamada, T., Wiker, H.G. and Harboe, M. 1988. Cloning and expression of the *Mycobacterium bovis* BCG gene for extracellular alpha antigen. J. Bacteriol. 170: 3847-3854.

Matsuo, T., Matsumoto, S., Ohara, N., Kitaura, H., Mizuno, A. and Yamada, T. 1995. Differential transcription of the MPB70 genes in two major groups of *Mycobacterium bovis* BCG substrains. Microbiology. 141: 1601-1607

Matsuo, T., Matsuo, H., Ohara, N., Matsumoto, S., Kitaura, H., Mizuno, A. and Yamada, T. 1996. Cloning and sequencing of an MPB70 homologue corresponding to MPB83 from *Mycobacterium bovis* BCG. Scand J. Immunol. 43: 483-489.

Matsuyama, S., Fujita, Y. and Mizushima, S. 1993. SecD is involved in the release of translocated secretory proteins from the cytoplasmic membrane of *Escherichia coli*. EMBO J. 12: 265-270.

Mattow, J., Jungblut, P.R., Muller, E.C. and Kaufmann, S.H. 2001. Identification of acidic, low molecular mass proteins of *Mycobacterium tuberculosis* strain H37Rv by matrix-assisted laser desorption/ionization and electrospray ionization mass spectrometry. Proteomics. 1: 494-507.

Mazmanian, S.K., Ton-That, H. and Schneewind, O. 2001. Sortase-catalysed anchoring of surface proteins to the cell wall of *Staphylococcus aureus*. Mol. Microbiol. 40: 1049-1057.

McVay, C.S. and Hamood, A.N. 1995. Toxin A secretion in *Pseudomonas aeruginas*: the role of the first 30 amino acids of the mature toxin. Mol. Gen. Genet. 249: 515-525.

Minamino, T. and Macnab, R.M. 1999. Components of the *Salmonella* flagellar export apparatus and classification of export substrates. J. Bacteriol. 181: 1388-1394.

Miura, K., Nagai, S., Kinomoto, M., Haga, S. and Tokunaga, T. 1983. Comparative studies with various substrains of *Mycobacterium bovis* BCG on the production of an antigenic protein, MPB70. Infect. Immun. 39: 540-545

Murphy, C.K. and Beckwith, J. 1994. Export of Proteins to the Cell Envelope in *Escherichia coli*. In: *Escherichia coli* and *Salmonella*: Cellular and Molecular Biology. F.C. Neidhardt, ed. ASM Press, Washington, DC. p.967-978

Nagai, S., Matsumoto, J. and Nagasuga, T. 1981b. Specific skin-reactive protein from culture filtrate of *Mycobacterium bovis* BCG. Infect. Immun. 31: 1152-1160.

Nagai, S., Wiker, H.G., Harboe, M. and Kinomoto, M. 1991. Isolation and partial characterization of major protein antigens in the culture fluid of *Mycobacterium tuberculosis*. Infect. Immun. 59: 372-382.

Nakano, H., Kawakami, Y. and Nishimura, H. 1992. Secretion of genetically-engineered dihydrofolate reductase from *Escherichia coli* using an *E. coli* alpha-hemolysin membrane translocation system. Appl. Microbiol. Biotechnol. 37: 765-771.

Orme, I.M. 1988. Induction of nonspecific acquired resistance and delayed-type hypersensitivity, but not specific acquired resistance in mice inoculated with killed mycobacterial vaccines. Infect. Immun. 56: 3310-3312.

Pallen, M. 2002. The ESAT-6/WXG100 superfamily- and a new Gram-positive secretion system? Trends Microbiol. 10: 209-212.

Palomaki, T. and Saarilahti, H.T. 1995. The extreme C-terminus is required for secretion of both the native polygalacturonase (PehA) and PehA-Bla hybrid proteins in *Erwinia carotovora* subsp. *carotovora*. Mol. Microbiol. 17: 449-459.

Palomaki, T. and Saarilahti, H.T. 1997. Isolation and characterization of new C-terminal substitution mutations affecting secretion of polygalacturonase in *Erwinia carotovora* ssp. *carotovora*. FEBS Lett. 400: 122-126.

Pathan, A.A., Wilkinson, K.A., Klenerman, P., McShane, H., Davidson, R.N., Pasvol, G., Hill, A.V. and Lalvani, A. 2001. Direct *ex vivo* analysis of antigen-specific IFN-gamma-secreting CD4 T cells in *Mycobacterium tuberculosis*-infected individuals: associations with clinical disease state and effect of treatment. J. Immunol. 167: 5217-5225.

Paulsen, I.T., Park, J.H., Choi, P.S. and Saier, M.H., Jr. 1997. A family of gram-negative bacterial outer membrane factors that function in the export of proteins, carbohydrates, drugs and heavy metals from gram-negative bacteria. FEMS Microbiol Lett. 156: 1-8.

Pollock, J., Douglas, A.J., Mackie, D.P. and Neill, S.D. 1994. Identification of bovine T-cell epitopes for three *Mycobacterium bovis* antigens: MPB70, 19,000 MW and MPB57. Immunology. 82: 9-15

Pollock, J.M., Girvin, R.M., Lightbody, K.A., Clements, R.A., Neill, S.D., Buddle, B.M. and Andersen, P. 2000. Assessment of defined antigens for the diagnosis of bovine tuberculosis in skin-test reactor cattle. Vet Rec. 146: 659-665

Portnoy, D.A., Moseley, S.L. and Falkow, S. 1981. Characterization of plasmids and plasmid-associated determinants of *Yersinia enterocolitica* pathogenesis. Infect. Immun. 31: 775-782.

Pugsley, A.P. 1993. The complete general secretory pathway in gram-negative bacteria. Microbiol Rev. 57: 50-108.

Quentin, Y., Fichant, G., and Denizot, F. 1999. Inventory, assembly and analysis of *Bacillus subtilis* ABC transport systems. J. Mol. Biol. 287: 467-484.

Randall, L.L., Hardy, S.J., Topping, T.B., Smith, V.F., Bruce, J.E. and Smith, R.D. 1998. The interaction between the chaperone SecB and its ligands: evidence for multiple subsites for binding. Protein Sci. 7: 2384-2390.

Roche, P.W., Triccas, J.A., Avery, D.T., Fifis, T., Billman-Jacobe, H. and Britton, S. 1994. Differential T cell responses to mycobacteria-secreted proteins distinguish vaccination with bacille Calmette-Guerin from infection with *Mycobacterium tuberculosis*. J. Infect. Dis. 170: 1326-1330.

Russel, M. 1998. Macromolecular assembly and secretion across the bacterial cell envelope: Type II protein secretion systems. J. Mol. Biol. 279:485-499.

Saleh, M.T., Fillon, M., Brennan, P.J. and Belisle, J.T. 2001. Identification of putative exported/secreted proteins in prokaryotic proteomes. Gene. 269: 195-204.

Salmond, G.P. and Reeves, P.J. 1993. Membrane traffic wardens and protein secretion in gram-negative bacteria. Trends Biochem Sci. 18: 7-12.

Sandkvist, M. 2001. Biology of type II secretion. Mol. Microbiol. 40: 271-283.

Schmidt, M.G. and Oliver, D.B. 1989. SecA protein autogenously represses its own translation during normal protein secretion in *Escherichia coli*. J. Bacteriol. 171: 643-649.

Schneider, G. 1999. How many potentially secreted proteins are contained in a bacterial genome? Gene. 237: 113-121.

Sharma, V., Sharma, S., Hoener zu Bentrup, K., McKinney, J.D., Russell, D.G., Jacobs, W.R., Jr. and Sacchettini, J.C. 2000. Structure of isocitrate lyase, a persistence factor of *Mycobacterium tuberculosis*. Nat. Struct. Biol. 7: 663-668.

Skjot, R.L., Oettinger, T., Rosenkrands, I., Ravn, P., Brock, I., Jacobsen, S. and Andersen, P. 2000. Comparative evaluation of low-molecular-mass proteins from *Mycobacterium tuberculosis* identifies members of the ESAT-6 family as immunodominant T-cell antigens. Infect. Immun. 68: 214-220.

Sonnenberg, M.G. and Belisle, J.T. 1997. Definition of *Mycobacterium tuberculosis* culture filtrate proteins by two-dimensional polyacrylamide gel electrophoresis, N-terminal amino acid sequencing, and electrospray mass spectrometry. Infect. Immun. 65: 4515-4524.

Sorensen, A.L., Nagai, S., Houen, G., Andersen, P. and Andersen, A.B. 1995. Purification and characterization of a low-molecular-mass T-cell antigen secreted by *Mycobacterium tuberculosis*. Infect. Immun. 63: 1710-1717.

Tekaia, F., Gordon, S.V., Garnier, T., Brosch, R., Barrell, B.G. and Cole, S.T. 1999. Analysis of the proteome of *Mycobacterium tuberculosis* in silico. Tuber. Lung Dis. 79: 329-342.

Terasaka, K., Yamaguchi, R., Matsuo, K., Yamazaki, A., Nagai, S. and Yamada, T. 1989. Complete nucleotide sequence of immunogenic protein MPB70 from *Mycobacterium bovis* BCG. FEMS Microbiol Lett. 49: 273-276.

Thanabalu, T., Koronakis, E., Hughes, C. and Koronakis, V. 1998. Substrate-induced assembly of a contiguous channel for protein export from *E. coli*: reversible bridging of an inner-membrane translocase to an outer membrane exit pore. EMBO J. 17: 6487-6496.

Tullius, M.V., Harth, G. and Horwitz, M.A. 2001. High extracellular levels of *Mycobacterium tuberculosis* glutamine synthetase and superoxide dismutase in actively growing cultures are due to high expression and extracellular stability rather than to a protein-specific export mechanism. Infect. Immun. 69: 6348-6363.

Ulstrup, J.C., Jeansson, S., Wiker, H.G. and Harboe, M. 1995. Relationship of secretion pattern and MPB70 homology with osteoblast-specific factor

2 to osteitis following *Mycobacterium bovis* BCG vaccination. Infect. Immun. 63: 672-675.

van Pinxteren, L.A., Ravn, P., Agger, E.M., Pollock, J. and Andersen, P. 2000. Diagnosis of tuberculosis based on the two specific antigens ESAT-6 and CFP10. Clin. Diagn. Lab. Immunol. 7: 155-160.

Vogel, J.P. and Isberg, R.R. 1999. Cell biology of *Legionella pneumophila*. Curr. Opin. Microbiol. 2: 30-34.

Vordermeier, H.M., Cockle, P.J., Whelan, A.O., Rhodes, S., Chambers, M.A., Clifford, D., Huygen, K., Tascon, R., Lowrie, D., Colston, M.J. and Hewinson, R.G. 2000. Effective DNA vaccination of cattle with the mycobacterial antigens MPB83 and MPB70 does not compromise the specificity of the comparative intradermal tuberculin skin test. Vaccine. 19: 1246-1255.

Vordermeier, H.M., Cockle, P.J., Whelan, A.O., Rhodes, S., Palmer, N., Bakker, D. and Hewinson, R.G. 1999. Development of diagnostic reagents to differentiate between *Mycobacterium bovis* BCG vaccination and *M. bovis* infection in cattle. Clin. Diagn. Lab. Immunol. 6: 675-682.

Vosloo, W., Tippoo, P., Hughes, J.E., Harriman, N., Emms, M., Beatty, D.W., Zappe, H. and Steyn, L.M. 1998. Characterization of a lipoprotein of *Mycobacterium bovis* (BCG) with sequence similarity to the secreted protein MPB70. Gene. 188: 123-128.

Wandersman, C., Binet, R., Letoffe, S., Ghigo, J.M. and Delepelaire, P. 1993. The general secretory pathway in bacteria. Trends Microbiol. 1: 249-250.

Wandersman, C. and Delepelaire, P. 1990. TolC, an *Escherichia coli* outer membrane protein required for hemolysin secretion. Proc. Natl. Acad. Sci. USA. 87: 4776-4780.

Weldingh, K., Rosenkrands, I., Jacobsen, S., Rasmussen, P.B., Elhay, M.J. and Andersen, P. 1998. Two-dimensional electrophoresis for analysis of *Mycobacterium tuberculosis* culture filtrate and purification and characterization of six novel proteins. Infect. Immun. 66: 3492-3500.

Weinrich Olsen, A., van Pinxteren, L.A., Meng Okkels, L., Birk Rasmussen, P., Andersen, P. 2001. Protection of mice with tuberculosis subunit vaccine based on a fusion protein of antigen 85B and ESAT-6. Infect. Immun. 69: 2773-2778.

Wiker, H.G., Harboe, M. and Lea, T.E. 1986. Purification and characterization of two protein antigens from the heterogeneous BCG85 complex in *Mycobacterium bovis* BCG. Int. Arch. Allergy Appl. Immunol. 81: 298-306.

Wiker, H.G. and Harboe, M. 1992. The antigen 85 complex: a major secretion product of *Mycobacterium tuberculosis*. Microbiol. Rev. 56: 648-661.

Wiker, H.G., Harboe, M. and Nagai, S. 1991. A localization index for distinction between extracellular and intracellular antigens of *Mycobacterium tuberculosis*. J. Gen. Microbiol. 137: 875-884.

Wiker, H.G., Lyashchenko, K., Aksoy, A.M., Lightbody, K.A., Pollock, J., Komissarenko, S.V., Bobrovnik, S.O., Kolesnikova, I.N., Mykhalsky, L.O., Gennaro, M.L. and Harboe, M. 1998. Immunochemical characterization of the MPB70/80 and MPB83 proteins of *Mycobacterium bovis*. Infect. Immun. 66: 1445-1452

Wiker, H.G., Michell, S.L., Hewinson, R.G., Spierings, E., Nagai, S. and Harboe, M. 1999. Cloning, expression and significance of MPT53 for identification of secreted proteins of *Mycobacterium tuberculosis*. Microb. Pathog. 26: 207-219.

Wiker, H.G., Nagai, S., Hewinson, R.G., Russell, W.P. and Harboe, M. 1996. Heterogenous expression of the related MPB70 and MPB83 proteins distinguish various substrains of *Mycobacterium bovis* BCG and *Mycobacterium tuberculosis* H37Rv. Scand. J. Immunol. 43: 374-380

Winstanley, C. and Hart, C.A. 2001. Type III secretion systems and pathogenicity islands. J. Med. Microbiol. 50: 116-126.

Young, D.B., Kaufmann, S.H., Hermans, P.W. and Thole, J.E. 1992. Mycobacterial protein antigens: a compilation. Mol. Microbiol. 6: 133-145.

From: *Genomics of GC-Rich Gram-Positive Bacteria*

7

TubercuList: a Regularly Updated Database Dedicated to the *Mycobacterium tuberculosis* Genome

Jean-Christophe Camus, Ivan Moszer, Louis M. Jones, Melinda J. Pryor, and Stewart T. Cole

Abstract

TubercuList is a public web server which provides access to a database for the *Mycobacterium tuberculosis* H37Rv genome and to an interface for users of the scientific community. For each predicted gene, a wealth of detailed biological and functional information is presented (DNA and proteins), including relevant bibliographic reference links. Many elaborate tools are available in order to help researchers obtain information, such as pattern searching and classic sequence analysis programs (BLAST, FASTA). Users can also view or download sequences, and consult the database using several search criteria, which can be easily combined. The data is regularly updated, and comments or requests from the users are scrupulously studied. This detailed and functional database is based on GenoList, a generic relational data model which integrates other complete bacterial genome data. New functionalities for multi-genome integration are in development. The URL of the TubercuList database is http://genolist.pasteur.fr/TubercuList/.

Introduction

Since the mid-nineties, the number of sequencing projects has increased exponentially, in particular for prokaryotic genomes. This significant progress corresponds to a desire from the scientific community to improve their knowledge in genomics, and is also due to the high-performance of new sequencing technologies and to the advent of new tools for biologists, e.g. transcriptomics and proteomics. One of the consequences is the number of biological databases and databanks has increased, these are generally public and available on the World-Wide Web. This information can be used by researchers in an effort to better understand the biochemistry, the physiological processes and the molecular biology of the cells. It can also assist in the identification and the understanding of some pathogenic mechanisms that are required in order to find new therapeutic drugs or vaccines for eradicating or preventing the propagation of diseases.

In this way, the complete genome of the *Mycobacterium tuberculosis* strain H37Rv was sequenced and annotated four years ago (Cole *et al.*, 1998). This human pathogenic bacterium is responsible for tuberculosis, one of the most important diseases in the world with 33% of the total population infected (http://www.who.int/inf-fs/en/fact104.html). This strain was first isolated in 1905 and it is the most used in tuberculosis research. Its genome is constituted of a circular chromosome of 4,411,532 base pairs, with an overall GC content of 65.6%. Its sequence has been recently re-annotated and contains 4045 potential coding sequences (CDS) to date.

This chapter describes in detail the corresponding database named TubercuList, available to the public on the http://genolist.pasteur.fr/TubercuList/ web server. This database contains biological and functional information about DNA and protein sequences, principally found in specialized literature or resulting from various *in silico* analyses. Many of the tools found in TubercuList are described in this text and allow the user to search for information or patterns in different ways.

Methods

Sequence Analysis And Annotation

The complete genome sequence of *Mycobacterium tuberculosis* strain H37Rv was obtained in 1998 by a combination of cosmid and whole-genome shotgun sequencing through a collaboration between the Sanger Centre (Hinxton, U. K.) and the Institut Pasteur (Paris, France) (Cole *et al.*, 1998). The sequence and annotation were deposited in EMBL/GenBank/DDBJ databanks (entry AL123456) and in the public database TubercuList presented here. With the aim of enhancing the utilization of this database, a completely updated version has recently been prepared. All the coding sequences (CDS) identified in 1998 have been re-evaluated and re-annotated using different tools, and several new CDS have also been included. During the first annotation as well as the present one, two independent approaches were used for identifying potential coding sequences. Firstly, the majority of the CDS were found manually, considering appropriate GC content and codon usage. Secondly, other CDS have been validated using information found in the literature or derived from personal communications; in this case, their localisation on the chromosome was determined using the TBLASTN program (Altschul *et al.*, 1990). Recently, a new program based on *M. tuberculosis* codon usage, named AMIGA, was also used to detect other possible CDS and sequence frameshifts (for details see Bocs *et al.*, 2002). This program identified potential CDS with a length >60 base pairs and merged the results with those generated by a modified GeneMark procedure; a coding probability was computed for each CDS. This set of CDS was then compared with those from the initial annotation, and only the additional genes predicted by the AMIGA program were investigated, based on database searches and criteria used in the manual approach.

For each of the coding sequences identified (manually or automatically), BLASTP (Altschul *et al.*, 1990) or FASTA (Pearson and Lipman, 1988) sequence comparison analysis - using non-redundant data from the EMBL, TrEMBL and SWISS-PROT databanks - were run and interpreted. In order to predict a function or to gain further information, complementary analyses have also been performed using the PROSITE database (Falquet *et al.*, 2002) and several web sites including information about enzymes (e.g. http://www.expasy.org/; http://srs.ebi.ac.uk/; http://www.worthington-biochem.com/; http://www.genome.ad.jp/kegg/kegg.html). The HTH (Helix-turn-helix) program was used to detect DNA binding motifs (Dodd and Egan, 1987; Dodd and Egan, 1990). Two others programs, TMHMM 1.0 (Sonnhammer *et al.*, 1998) and SignalP (Nielsen *et al.*, 1997), were also used to predict the subcellular localisation of putative gene products. As the transport of many compounds is essential, particularly for pathogenic bacteria which must survive in a hostile environment of host cells, important studies have also been carried out on proteins involved in transport, using notably the recent work of J. Content and I. Paulsen (Braibant *et al.*, 2000; http://

www.biology.ucsd.edu/~ipaulsen/transport/; Paulsen *et al.*, 2000). In addition, a comparative analysis of several predicted products classified in *M. tuberculosis* partitions or protein families has been done using the MEME/ MAST programs (Bailey and Elkan, 1994; Bailey and Gribskov, 1998) as previously described in Tekaia *et al.* (1999).

In the first stage, the work on annotations was supported by the Artemis software (http://www.sanger.ac.uk; Rutherford *et al.*, 2000) and, in the second stage, resulting information was transferred to TubercuList as described below.

The GenoList Model

The TubercuList database was constructed using a generic database model called GenoList. GenoList derives from the work performed on the Colibri (Médigue *et al.*, 1993) and SubtiList (Moszer *et al.*, 1995; Moszer *et al.*, 2002) databases, dedicated to the genomes of *Escherichia coli* and *Bacillus subtilis* respectively. This model has evolved from a parceled view of bacterial genomes (before 1997) towards an original data structure designed for the representation, manipulation, and maintenance of complete microbial genomes. The internal organization of the data involves artificial sub-divisions of the genome sequence. Associated genomic features are located relative to these chromosomal sections, which are then mapped to the whole chromosome. This leads to partial and efficient loading of the data whenever needed (query and update). Thus sequence and feature updates do not require re-evaluation of the whole set of objects in the database, only of the chromosome fragment concerned. This artifice is completely transparent to the end-user of the database, through the user-friendly WWW interface described below. The conceptual data schema was implemented as a relational database using the Sybase® SQL DataBase Management System.

A user-oriented interface for browsing and querying the database was developed on top of the data model (see next sections). It is accessible on the Internet through a web server which runs dynamic HTML pages, generated by a CGI script written with the C/C++ programming languages. The interface is aimed at providing the user with facilitated access to the data he is interested in, according to the most frequent questions asked by researchers.

Data are transferred from the Artemis software to TubercuList through the EMBL file format, which is parsed by Perl scripts to generate the SQL statements required to populate the database. Other sources of informations can be imported as well, using dedicated procedures.

The TubercuList Web Interface

The TubercuList web page is divided in three frames described in detail in this part. The user can change the vertical size of the two frames on the right (Figures 1 and 2).

The Query Frame (Vertical Left)

This frame gives straight access to the main queries the biologist is interested in (Figures 1 and 2). First, the user can perform a query by “Gene name” by filling in the corresponding text field. Three options allow the user to quickly obtain further information derived from the gene name filled in. With the “Partial name” option, the user can access all genes whose name begins with the provided characters. The “Synonym” option allows one to find genes whose possible alternative names are stored as synonyms. The “Region” option is the most useful one as it provides immediate access to the chromosome region (default 20 kb, this value can be changed) around the gene of interest, either as a gene list or as a graphical representation (see below).

The “Location” query allows the user to display a chromosome region according to two coordinates (“From” and “To”, or “From” only to get a 20 kb long region around this position).

“Free text” can be used to search for a character string (not necessarily a word) in all the description fields of the database related to genes (see below the description of the lower right frame).

The “Functional category” query allows the user either to display a hierachical clickable tree of the functional categories that can be used to retrieve genes assigned to each (sub-)category, or to directly type in a code corresponding to a known category to get the corresponding genes.

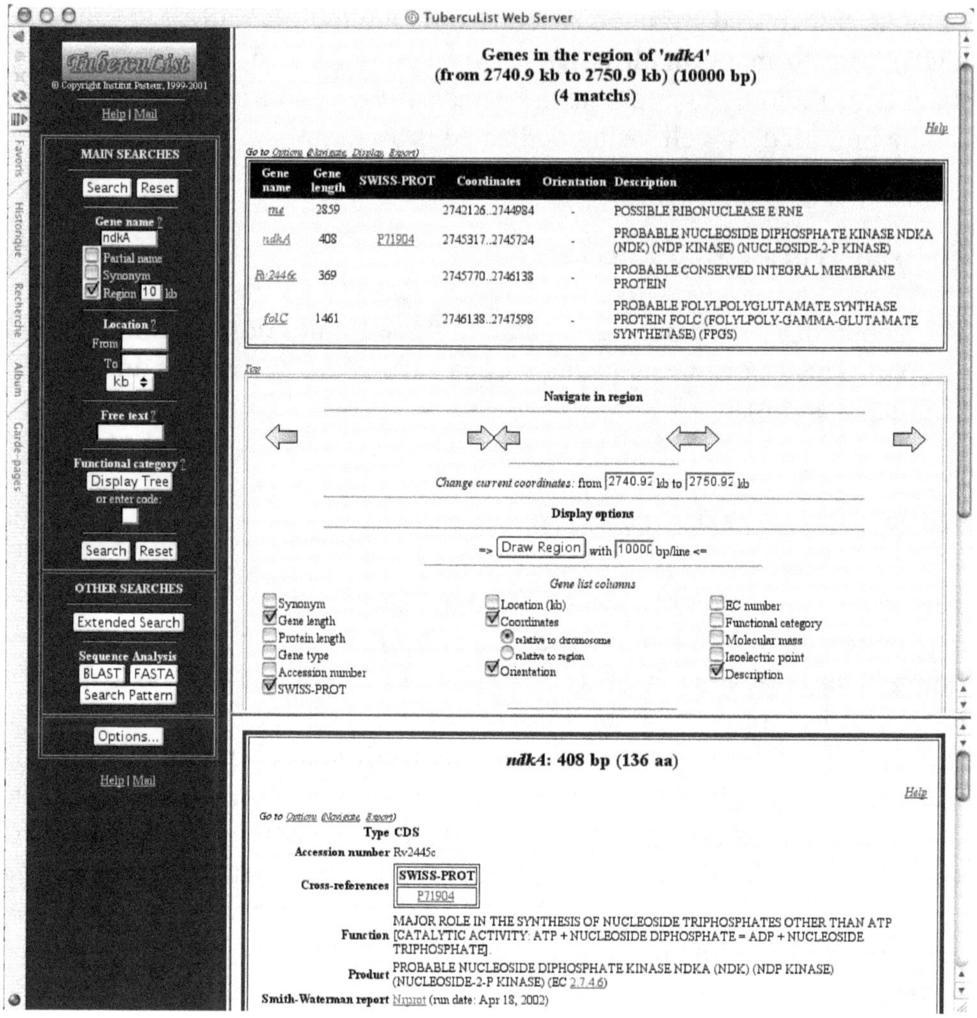

Figure 1. TubercuList web page with the *ndkA* genomic region in list format and a detailed *ndkA* gene annotation (first part).

The "Extended Search" button displays a new form in the upper right frame that gives the possibility to define several search criteria based on all the database fields, which can also be combined, in order to ask complex questions to the database. This functionality is directly derived from the relational capabilities of the underlying database model.

Sequence analysis tools are readily available. The "BLAST" and "FASTA" buttons allow the user to launch the respective algorithms (various flavours available) to search for sequence similarities, using a user-provided query sequence against the DNA or protein sequences of *M. tuberculosis*. The

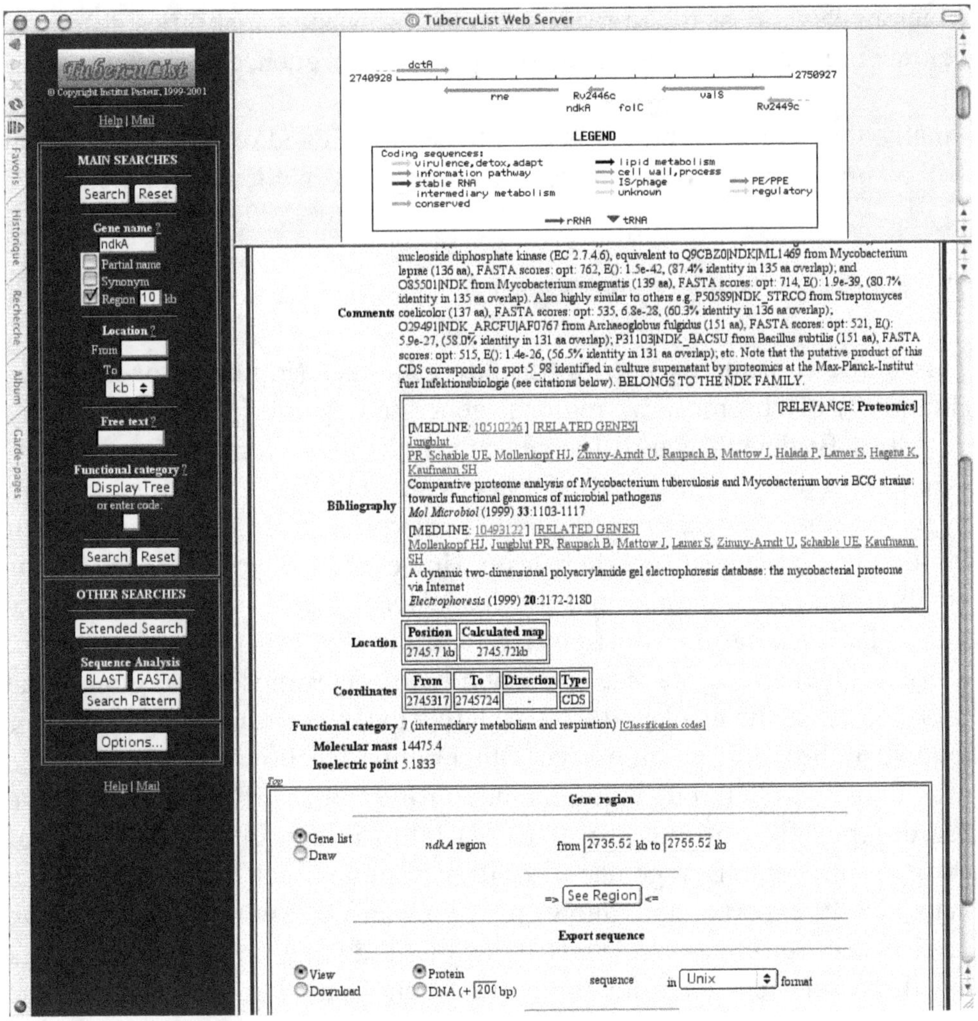

Figure 2. TubercuList web page with the *ndkA* genomic region in drawing format and a detailed *ndkA* gene annotation (second part).

"Search Pattern" function is particularly useful for biologists: it allows them to search for degenerate DNA/protein pattern sequences in the genome/proteome of *M. tuberculosis*, taking into account constraints such as the location of the requested pattern occurrences with respect to the gene boundaries. The syntax used to describe the patterns allows one to define ambiguous or mandatory positions, the maximum number of mismatches allowed, and complex patterns made up of several small patterns separated by a variable number of letters. In both cases (BLAST/FASTA and pattern searches), results of sequence searches are presented together with the relevant genomic annotations (e.g. genes located downstream the pattern occurences), which makes it possible to readily use the browsing capabilities described below in relation with the results of a particular sequence analysis.

Finally, the "Options" button is used to set search and display options that will be preserved from session to session on a given client computer.

The Gene List Frame (Upper Right)

This frame is primarily used to display gene lists (Figure 1) or drawings (Figure 2) of chromosome regions, generated by the queries previously described. By default, gene lists contain the following columns: "Gene name", "Gene length", "SWISS-PROT" accession number (with the relevant hypertext link), "Location (kb)", and "Description" (see the explanation of these fields in the next section below). However the layout of these lists is highly customizable through an option panel located beneath the list. Columns to be displayed can be (de-)selected, and the sort order can be changed. Moreover, when a gene list corresponds to a chromosome region, gene coordinates can be displayed either relative to the corresponding region, or relative to the whole chromosome. Other options (yellow arrows) allow the user to easily walk through the genome starting from a given location. These region-specific options are also available under graphical maps of chromosome regions, together with further options such as the scale of the map. Finally, export options allow one to view in a new window or to retrieve in a local file information related to gene lists/drawings: tab-delimited tables of all the selected fields for the current gene selection, direct or reverse DNA sequence in the case of a chromosome region, gene sequences (DNA, RNA, or protein) in either cases.

This frame is also used to display the clickable functional tree, the extended search form, and the forms and results of sequence analysis tools (BLAST/FASTA, pattern search) (see above).

The Gene Detail Frame (Lower Right)

In this frame, the user can find a detailed annotation about the protein or RNA coding sequences selected in the upper frame. These annotations contain some or all of the possible fields described below.

- Accession number (Figure 1): e.g. Rv0470c
  It is generally a Rv_number, which is unique and has been attributed sequentially to each protein coding sequence. The letter c at the end indicates that the potential CDS is located on the complementary strand (direction "-").

  When new CDS have been identified, the numbering of the new coding sequences has not interfered with the labelling of the existing genes. Indeed a new CDS uses the same Rv_number as the preceding gene followed by a letter extension never used before (e.g. Rv2307A, Rv2307B, and Rv2307D have been introduced between Rv2307c and Rv2308).

  Note that the the initiation codon for Rv0001, predicted to encode the DnaA product, was chosen as the start point for the chromosome sequence numbering.

- Gene name (Figure 1): e.g. *pcaA*
  When a gene name is attributed to a CDS, it generally comes from consistent information found in the literature (with attribution classified by priority: *M. tuberculosis* > other mycobacteria > *Escherichia coli* > *Bacillus subtilis* > other organisms).

- Type (Figure 1): e.g. tRNA
  Indicates the type of the selected gene, e.g. CDS, rRNA or tRNA.

- Length (Figure 1): e.g. 1521 bp (507 aa)
  Corresponds to the size of the CDS (in base pairs) and the length of the putative product (in amino acids) when thought to encode a protein.

- Cross-references (Figure 1): e.g. SWISS-PROT P49993
  If the selected protein coding sequence has a cross-reference in the SWISS-PROT databank, a link in TubercuList has been created towards the corresponding page in the ExPASy web site.

- Location (Figure 2): e.g. Position 302.9 kb Calculated map 302.87kb
  Coordinates: e.g. From 302866 To 305424 Direction + Type CDS
  Corresponds to the positions of the selected gene on the chromosome (direction "-" if the gene is transcribed on the complementary strand). The coordinates are in nucleotides.

- Product (Figure 1): e.g. PROBABLE SUCCINYL-COA:3-KETO-ACID-COENZYME A TRANSFERASE (BETA SUBUNIT) SCOB

(3-OXO-ACID:COA TRANSFERASE) (EC 2.8.3.5)
In this field, the user can find the main name of the product (e.g. 3-DEHYDROQUINATE DEHYDRATASE), followed in parentheses by the other known names (e.g. 3-DEHYDROQUINASE and TYPE II DHQASE).

A multifunctional protein is annotated as follows:
POSSIBLE BIFUNCTIONAL ENZYME RIBOFLAVIN BIOSYNTHESIS PROTEIN RIBD: DIAMINOHYDROXY-PHOSPHORIBOSYLAMINOPYRIMIDINE DEAMINASE (RIBOFLAVIN-SPECIFIC DEAMINASE) + 5-AMINO-6-(5-PHOSPHORIBOSYL-AMINO) URACIL REDUCTASE (HTP REDUCTASE) (EC 3.5.4.26; 1.1.1.193).

When necessary, a characterised chain or subunit is also indicated in parentheses. In addition, a tag (PROBABLE, POSSIBLE or PUTATIVE) written before the main product name gives an indication about the level of confidence. If the product has been experimentally characterised and functionally studied in *M. tuberculosis*, none of these tags was added.

This part of "Product" corresponds to the information found in the "Description" column, in the upper frame (see above).

In addition, when a putative product is thought to be an enzyme, the user can find the EC (Enzyme Commission) number(s) related to the Nomenclature Committee of the International Union of Biochemistry and Molecular Biology (NC-IUBMB), with the relevant hypertext link(s) to the ExPASy server when applicable. Different sources have been used to complete this enzymatic information (see Methods).

- Function (Figure 1): e.g. INVOLVED IN FATTY ACID SYNTHESIS [CATALYTIC ACTIVITY: SUCCINYL-COA + A 3-OXO ACID = SUCCINATE + A 3-OXO-ACYL-COA].
  When a protein coding sequence has been selected, the user can read a short description of the function. This information is generally extracted from the literature or from the field "FUNCTION" in the "Comments" section of the SWISS-PROT databank. Whenever possible, the catalytic activity of the putative enzyme has been mentioned.

- Smith-Waterman report (Figure 1): e.g. Nrprot (run date: Apr 18, 2002)
  When the user clicks on this link, a new window appears with sequence similarity results from a Smith-Waterman search ran against the NCBI non-redundant protein databank. These reports are regularly updated (the date in parentheses corresponds to the date of the last run).

- Comments (Figure 2):
  This field contains a lot of other information collected during the annotation process. Users can find a description of the putative product (including its length and its name(s)), the main FASTA or BLAST scores (similarity with products from *M. leprae* and other mycobacteria, then from other organisms), PROSITE or helix-turn-helix domains identified in the corresponding protein, and the protein family name. During the annotation, the amino acid distribution of each gene product was also studied and results may be included here. In addition, when the length or the gene name has been altered (resulting from a new annotation), changes are indicated and the old name is retained. Most proteins which have been identified in culture supernatants or in the cell wall by proteomic studies are also tagged (Weldingh *et al.*, 1998; Jungblut *et al.*, 1999; Mollenkopf *et al.*, 1999; Rosenkrands *et al.*, 2000a; Rosenkrands *et al.*, 2000b).

- Bibliography (Figure 2):
  Some genes are linked to targeted citation(s) which correspond to the most important publications related to the gene of interest. A relevance value indicates the reason for which a particular reference was linked to a given gene (e.g. its function, its regulation, its protein structure, etc.).

- Molecular mass (Figure 2): e.g. 56550.7
  This number is the molecular weight of the predicted product, in Dalton, calculated by the pepstats program.

- Isoelectric point (Figure 2): e.g. 5.3951
  pI of the corresponding protein, estimated with the pepstats program.

- Functional category (Figure 2): e.g. 3 (cell wall and cell processes)
  Genes have been classified into two functional classification schemes. The first one corresponds to the eleven categories listed in Table 1. Note that class 6 (PE/PPE) corresponds to proteins belonging to one of the two large unrelated families of acidic, glycine-rich proteins, the PE and PPE families, specific for mycobacteria (for details, see Cole *et al.*, 1998). The second classification is more detailed and is adapted from the work of M. Riley (Riley, 1993).

  For each kind of classification, a link gives access to the whole classification list.

Table 1. Functional classification used on TubercuList

Number	Functional class	Color in map
0	Virulence, detoxification, adaptation	white
1	Lipid metabolism	black
2	Information pathway	red
3	Cell wall and cell process	green
4	Stable RNAs	blue
5	Insertion sequences and phages	turquoise
6	PE/PPE	purple
7	Intermediary metabolism and respiration	yellow
8	Unknown	light green
9	Regulatory proteins	light blue
10	Conserved hypotheticals	amber

- MAST: e.g. P8.2
  When the putative product of a CDS can be classified in a partition or a protein family (Tekaia *et al.*, 1999), results of partition analysis can be consulted. If the user clicks on the corresponding link, he can see results generated by the MEME (Multiple Expectation maximization for Motif Elicitation) or MAST (Motif Alignment and Search Tool) programs, with three sections: HIGH-SCORING SEQUENCES, MOTIF DIAGRAMS and ANNOTATED SEQUENCES.

  Another link on the right ([Partition Functions]) opens a new window containing a list of all the *M. tuberculosis* partitions and the predicted putative functions (UNK for unknown).

- TMHMM: e.g. Rv2920c
  If the TMHMM program detects potential hydrophobic segment(s) in a predicted protein, a link is added on TubercuList, and the score(s) and a diagram with position(s) can be seen in the window created.

Other options in this frame allow the user to access data about genetic and physical mapping information. Thus, the user can view the chromosomal region around the gene of interest using options in the "Gene region" section (drawing or gene list). Default values for the limits of the region correspond to a span of 10 kb on each side of the gene, but the user can change these parameters.

Table 2. General features of the *M. tuberculosis* H37Rv genome

Genome feature	Number
Chromosome length (bp)	4,411,532
Overall GC content (%)	65,61
Protein coding genes	3,995
tRNA	45
Other RNA	5
Gene density (bases per gene)	1,104
Average gene length (bases per gene)	1,004
CDS with function assigned	2,052
Unknown CDS without homology	376
Gene name attributed	>1500
Potential coding capacity (%)	90.9
PE/PPE sequences (%)	9
Insertion sequences and phage sequence (%)	3.4

In the "Export sequence" section, the user can also view the DNA sequence (or the protein sequence when applicable), and can download data in a file. It is possible to select the file format: Unix, Macintosh, or Windows.

Current Release Of TubercuList (R4)

The latest version of the TubercuList database contains 3995 potential protein coding sequences, and 50 genes coding for stable RNA: 3 rRNA (5S, 16S, 23S), the 10Sa RNA (*ssr*), the RNA component of ribonuclease P (*rnpB*), and 45 tRNA. The general features of the *M. tuberculosis* genome are shown in Table 2.

Gene descriptions contain detailed information about the function assigned, several external links (e.g. cross-references and bibliographic references), and the different fields are homogeneously annotated. Numerous and major changes occured as compared to the first release of TubercuList in 1998 (Table 3). All the predicted genes are classified into one of the eleven functional category listed in Table 1 and Figure 3 shows this repartition in 1998 and in the current release R4. The classification of more than 600 CDS identified in 1998 has been changed. Not unexpectedly, most of these changes corresponds to transfers of CDS originally classified in the classes 8 (unknown proteins) or 10 (conserved hypothetical proteins). Thus 94

Table 3. Summary of the main modifications occured in TubercuList as compared to the first release

Changes	Number
Nucleotide corrected	4
Gene length extended	16
Gene length shortened	31
CDS added	82
CDS removed	10
Rv_number -> gene name	307
Gene name -> other gene name	64
Gene name -> Rv_number	24
Functional class attribution changed[#]	621
CDS updated without affecting the classification[#]	331

[#] classification of the eleven categories reported in Table 1.

functions have been attributed to CDS previously assigned to class 8. Approximatively 350 genomic annotations have also been significantly altered without a transfer to another class. Almost 400 gene names have changed (attribution, loss or modification). All the re-annotations were principally based on new significant sequence similarities with other proteins or experimental data from pertinent publications (studies on *M. tuberculosis* or other organisms). New information also came from personal communications. Analysis of synteny information, in particular with *Mycobacterium leprae*, has also been used for updating TubercuList. The length of a few genes has also been updated, based on the results of BLASTP/FASTA and MEME/MAST analysis, or due to correction of the DNA sequence (Table 3).

In addition, 82 new protein coding genes have been included since the first release (a few of them have been experimentaly characterized by proteomic analysis), and 10 genes have been removed (9 new genes replaced CDS previously located on the opposite strand).

In total, 25% of the putative genes identified in 1998 have had their annotation significantly altered. All the protein coding genes now contain the "Product" and "Function" fields. The "Comments" field of all the CDS has been systematically rewritten for a better homogeneity in annotation, and more than 600 links for relevant bibliographic references have been added in the database.

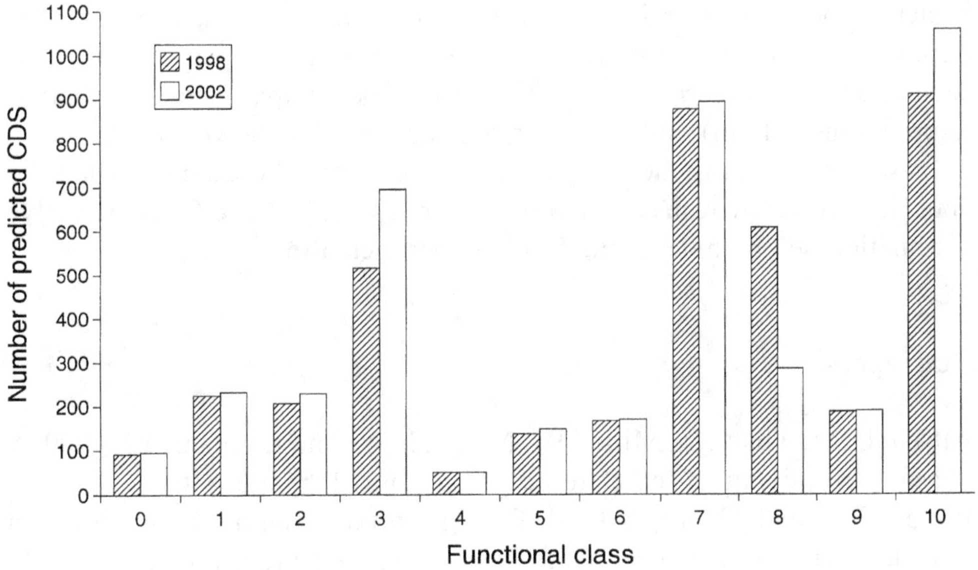

Figure 3. Functional distribution of the *M. tuberculosis* genes in 1998 and 2002. For the definitions of the different classes, refer to Table 1.

To date, TubercuList contains 2,052 proteins with assigned functions (52% of the 3,995 proteins predicted, vs 40% in 1998), and only 376 putative proteins share no homology with other known proteins; some of these may be specific to *M. tuberculosis*.

The current release of TubercuList also contains an updated list of all the MIRUs (Mycobacterial Intergenic Repetitive Units) inventoried in the genome, with their positions and their related class.

Conclusions

TubercuList is an important relational database, easy to use, which allows the user to find pertinent information about the *M. tuberculosis* genome using various search criteria, and also to utilize elaborate sequence analysis tools. Owing to the fact that sequencing projects and biological databases are in a constant progression, this server must be regularly updated and developed, trying to continuously satisfy the users. Its purpose is to collate and integrate various aspects of the genomic information from *M. tuberculosis*, but also from the other members of the *M. tuberculosis* complex: *M. africanum*, *M. bovis*, *M. bovis* BCG, *M. microti*, and *M. canetti*. TubercuList has been constructed based on the GenoList model, and other databases for other bacterial species can be integrated into the same data structure. Thus the Leproma web server contains full genomic information from the recently sequenced *Mycobacterium leprae* (Cole *et al.*, 2001; http://genolist.pasteur.fr/Leproma/index.html), and a new database dedicated to the *M. bovis* genome, also using the GenoList model, will be available soon. These similar databases could help researchers in comparative genomics analysis and for constructive speculation before undertaking future experimentation.

References

Altschul, S.F., Gish, W., Miller, W., Myers, E.W., and Lipman, D.J. 1990. A basic local alignment search tool. J. Mol. Biol. 215: 403-410.

Bailey, T.L., and Elkan, C. 1994. Fitting a mixture model by expectation maximization to discover motifs in biopolymers. Proc. Int. Conf. Intell. Syst. Mol. Biol. 28-36.

Bailey, T.L., and Gribskov, M. 1998. Combining evidence using p-values: application to sequence homology searches. Bioinformatics. 14: 48-54.

Bocs, S., Danchin, A. and Médigue, C. 2002. Re-annotation of genome microbial CoDing-Sequences: finding new genes and inaccurately annotated genes. BMC Bioinformatics. 3: 5.

Braibant, M., Gilot, P., and Content, J. 2000. The ATP binding cassette (ABC) transport systems of *Mycobacterium tuberculosis*. FEMS Microbiol. Rev. 24: 449-467.

Cole, S.T., Brosch, R., Parkhill, J., Garnier, T., Churcher, C., Harris, D., Gordon, S.V., Eiglmeier, K., Gas, S., Barry, C.E. 3rd, Tekaia, F., Badcock, K., Basham, D., Brown, D., Chillingworth, T., Connor, R., Davies, R., Devlin, K., Feltwell, T., Gentles, S., Hamlin, N., Holroyd, S., Hornsby, T., Jagels, K., Krogh, A., McLean, J., Moule, S., Murphy, L., Oliver, K., Osborne, J., Quail, M.A., Rajandream, M.-A., Rogers, J., Rutter, S., Seeger, K., Skelton, J., Squares, R., Squares, S., Sulston, J.E., Taylor, K., Whitehead, S., and Barrell, B.G. 1998. Deciphering the biology of *Mycobacterium tuberculosis* from the complete genome sequence. Nature. 393: 537-544.

Cole, S.T., Eiglmeier, K., Parkhill, J., James, K.D., Thomson, N.R., Wheeler, P.R., Honore, N., Garnier, T., Churcher, C., Harris, D., Mungall, K., Basham, D., Brown, D., Chillingworth, T., Connor, R., Davies, R.M., Devlin, K., Duthoy, S., Feltwell, T., Fraser, A., Hamlin, N., Holroyd, S., Hornsby, T., Jagels, K., Lacroix, C., Maclean, J., Moule, S., Murphy, L., Oliver, K., Quail, M.A., Rajandream, M.-A., Rutherford, K.M., Rutter, S., Seeger, K., Simon, S., Simmonds, M., Skelton, J., Squares, R., Squares, S., Stevens, K., Taylor, K., Whitehead, S., Woodward, J.R., and Barrell, B.G. 2001.

Massive gene decay in the leprosy bacillus. Nature. 409: 1007-1011.

Dodd, I.B., and Egan, J.B. 1987. Systematic method for the detection of potential lambda cro-like DNA-binding regions in proteins. J. Mol. Biol. 194: 557-564.

Dodd, I.B., and Egan, J.B. 1990. Improved detection of helix-turn-helix DNA-binding motifs in protein sequences. Nucleic Acids Res. 18: 5019-5026.

Falquet, L., Pagni, M., Bucher, P., Hulo, N., Sigrist, C.J., Hofmann, K., and Bairoch, A. 2002. The PROSITE database, its status in 2002. Nucleic Acids Res. 30: 235-238.

Jungblut, P.R., Schaible, U.E., Mollenkopf, H.J., Zimny-Arndt, U., Raupach, B., Mattow, J., Halada, P., Lamer, S., Hagens, K., and Kaufmann, S.H. 1999. Comparative proteome analysis of *Mycobacterium tuberculosis* and *Mycobacterium bovis* BCG strains: towards functional genomics of microbial pathogens. Mol. Microbiol. 33: 1103-1117.

Médigue, C., Viari, A., Hénaut, A., and Danchin, A. 1993. Colibri: a functional data base for the *Escherichia coli* genome. Microbiol. Rev. 57: 623-654.

Mollenkopf, H.J., Jungblut, P.R., Raupach, B., Mattow, J., Lamer, S., Zimny-Arndt, U., Schaible, U.E., and Kaufmann, S.H. 1999. A dynamic two-dimensional polyacrylamide gel electrophoresis database: the mycobacterial proteome via Internet. Electrophoresis. 20: 2172-2180.

Moszer, I., Glaser, P., and Danchin, A. 1995. SubtiList: a relational database for the *Bacillus subtilis* genome. Microbiology. 141: 261-268.

Moszer, I., Jones, L.M., Moreira, S., Fabry, C., and Danchin, A. 2002. SubtiList: the reference database for the *Bacillus subtilis* genome. Nucleic Acids Res. 30: 62-65.

Nielsen, H., Engelbrecht, J., Brunak, S., and von Heijne, G. 1997. Identification of prokaryotic and eukaryotic signal peptides and prediction of their cleavage sites. Protein Eng. 10: 1-6.

Paulsen, I.T., Nguyen, L., Sliwinski, M.K., Rabus, R., and Saier, M.H. Jr. 2000. Microbial genome analyses: comparative transport capabilities in eighteen prokaryotes. J. Mol. Biol. 301: 75-100.

Pearson, W.R., and Lipman, D.J. 1988. Improved tools for biological sequence comparison. Proc. Natl. Acad. Sci. USA. 85: 2444-2448.

Riley, M. 1993. Functions of the gene products of *Escherichia coli*. Microbiol. Rev. 57: 862-952.

Rosenkrands, I., King, A., Weldingh, K., Moniatte, M., Moertz, E., and Andersen, P. 2000. Towards the proteome of *Mycobacterium tuberculosis*. Electrophoresis. 21: 3740-3756.

Rosenkrands, I., Weldingh, K., Jacobsen, S., Hansen, C.V., Florio, W., Gianetri, I., and Andersen, P. 2000. Mapping and identification of *Mycobacterium tuberculosis* proteins by two-dimensional gel electrophoresis, microsequencing and immunodetection. Electrophoresis. 21: 935-948.

Rutherford, K., Parkhill, J., Crook, J., Horsnell, T., Rice, P., Rajandream, M.-A., and Barrell, B. 2000. Artemis: sequence visualization and annotation. Bioinformatics. 16: 944-945.

Sonnhammer, E.L., von Heijne, G., and Krogh, A. 1998. A hidden Markov model for predicting transmembrane helices in protein sequences. Proc. Int. Conf. Intell. Syst. Mol. Biol. 6: 175-182.

Tekaia, F., Gordon, S.V., Garnier, T., Brosch, R., Barrell, B.G., and Cole, S.T. 1999. Analysis of the proteome of *Mycobacterium tuberculosis in silico*. Tuber. Lung Dis. 79: 329-342.

Weldingh, K., Rosenkrands, I., Jacobsen, S., Rasmussen, P.B., Elhay, M.J., and Andersen, P. 1998. Two-dimensional electrophoresis for analysis of *Mycobacterium tuberculosis* culture filtrate and purification and characterization of six novel proteins. Infect. Immun. 66: 3492-3500.

Index

www.ingramcontent.com/pod-product-compliance
Lightning Source LLC
LaVergne TN
LVHW080057220926
848484LV00007BA/1783